CW01081905

Q.E.D.

First published in the United States of America in 2004 by
Walker Publishing Company, Inc.

Published simultaneously in Canada by Fitzhenry and Whiteside,
Markham, Ontario L3R 4T8

For information about permission to reproduce selections from this book, write to Permissions, Walker & Company,
104 Fifth Avenue, New York, New York 10011.

Library of Congress Cataloging-in-Publication Data

Polster, Burkard.

Q.E.D.: beauty in mathematical proof / written and illustrated by Burkard Polster.

p. cm.

ISBN 0-8027-1431-5 (alk. paper)

1. Proof theory. 2. Logic, Symbolic and mathematical. I. Title: QED: beauty in mathematical proof. II. Title.

QA9.54.P65 2004
511.3'6—dc22 2004041905

Visit Walker & Company's Web site at www.walkerbooks.com

Printed in the United States of America

2 4 6 8 10 9 7 5 3 1

Q.E.D.

BEAUTY IN MATHEMATICAL PROOF

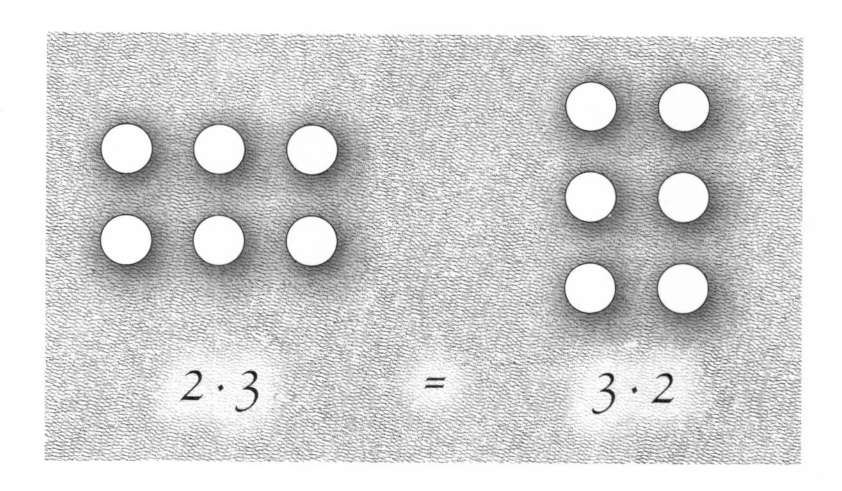

written and illustrated by

Burkard Polster

WOODEN BOOKS

Walker & Company
New York

With love to Anu, who understands it all . . .

I am indebted to the many mathematicians of the past and present from whose ideas this book has been distilled. I am grateful to Marty Ross and John Stillwell for their criticism and insightful comments. Finally, many, many thanks to John Martineau and Daud Sutton for being patient guides and accomplices in opening this visual vortex into the beautiful world of mathematical proofs.

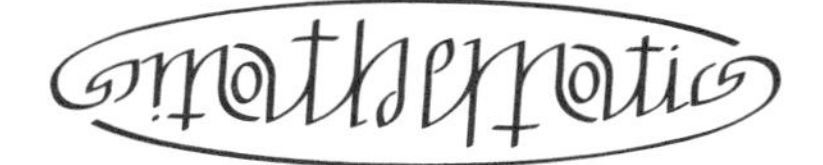

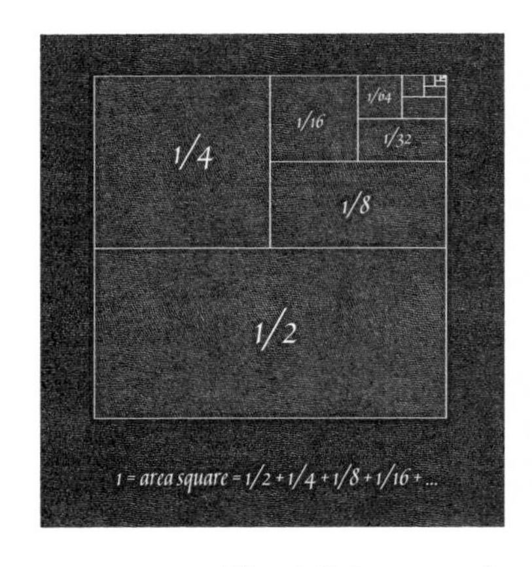

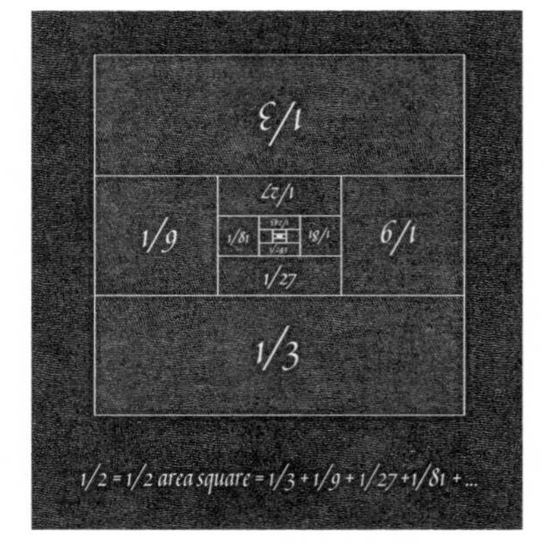

Two infinite sums—boxed and ready to be served.

Contents

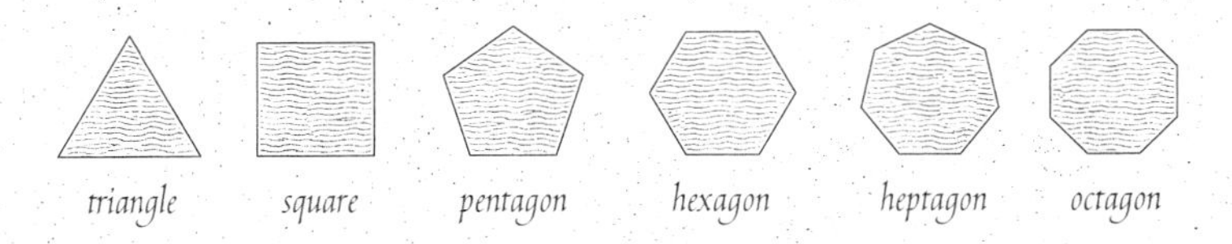

A regular polygon is a convex figure with equal sides and angles. There are infinitely many regular polygons.

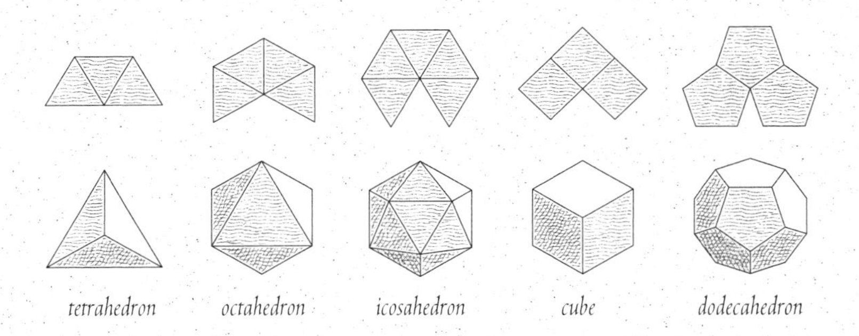

A regular polyhedron is a convex body with identical regular polygons as faces and the same number of faces meeting at every corner. Shown at the top are the different ways of joining three or more identical regular polygons to a corner with space left to fold up into three dimensions. These possibilities of building spatial corners then can be shown to extend in a unique way to the famous five regular polyhedra.

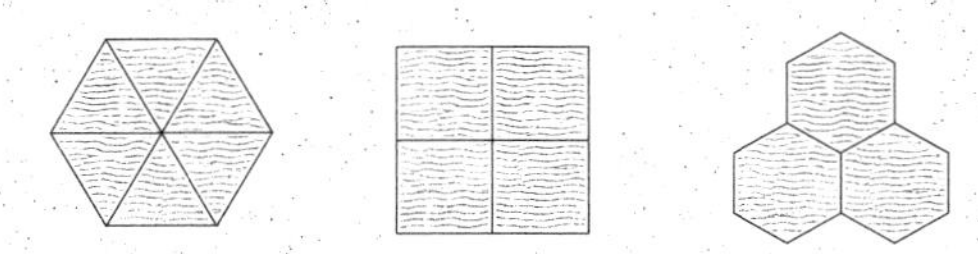

The same simple reasoning shows that there are three tilings of the plane with identical regular polygons.

INTRODUCTION

There are some mathematical objects whose beauty everyone is able to appreciate. The regular polygons and polyhedra are good examples—these figures are surpassed in perfection only by the circle and the sphere. Then there is Pythagoras's theorem, a cornerstone of the right-angled worlds we build for ourselves, and perhaps the conic sections, which describe the orbits of celestial bodies.

Very few people appreciate more than some elementary aspects of mathematical beauty, much of it revealing itself only to mathematicians in the study and creation of intricately crafted proofs, barely within the reach of the most highly trained human minds.

As a mathematician, I declare that I have established the truth of a theorem by writing at the end of its proof the three letters *Q.E.D.*, an abbreviation for the Latin phrase *quod erat demonstrandum*, which translates as "what had to be proved." On the one hand, Q.E.D. is a synonym for truth and beauty in mathematics; on the other hand, it represents the seemingly inaccessible side of this ancient science.

Q.E.D. can, however, also be found at the end of some simple, striking, and visually appealing proofs. This little book presents a journey through a collection of these wondrous gems, exploring the ideas behind mathematical proof on the way, written for all those who are interested in the beauty of mathematics hidden below the surface.

Melbourne, July 2003

TREACHEROUS TRUTH

what proofs are all about

In mathematics, as in the physical sciences, we may run an experiment or check a few cases to come up with a conjecture for a theorem. However, in mathematics experiments cannot replace proof, no matter how natural and obvious the conjecture is that they support. For example, the maximum number of regions defined by 1, 2, 3, 4, 5, and 6 points on a circle (*below*) is 1, 2, 4, 8, 16, and . . . 31, not 32!

Or, take the famous Goldbach conjecture, which claims that every even number greater than two is the sum of two prime numbers as, for example, $12 = 5 + 7$ or $30 = 23 + 7$. Although this conjecture has been checked for many millions of cases, unless a proof is found, we cannot be sure that the next case we check won't show that the conjecture is false.

Proofs should be as short, transparent, elegant, and insightful as possible. Our proof (*opposite, top*) that the number 0.999 . . . (with infinitely many 9s) is equal to 1 is of this kind, and its main argument can be easily adapted to convert any decimal number with one of those slightly worrysome infinitely repeating tails into a number we are more comfortable with. The proof that the indented chessboard cannot be tiled with dominoes (*opposite, center*) is another example. Of course, the argument here applies to many other mutilated chessboards.

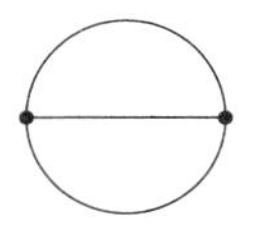

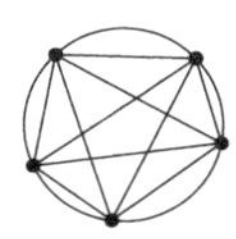

Theorem: $1 = 0.999\ldots$

Proof: Set $x = 0.999\ldots$. Then

$$\begin{array}{rl} 10x &= 9.999\ldots \\ - \quad x &= 0.999\ldots \\ \hline = \quad 9x &= 9.000\ldots \end{array}$$

Thus $x = 1.000\ldots$

Q.E.D.

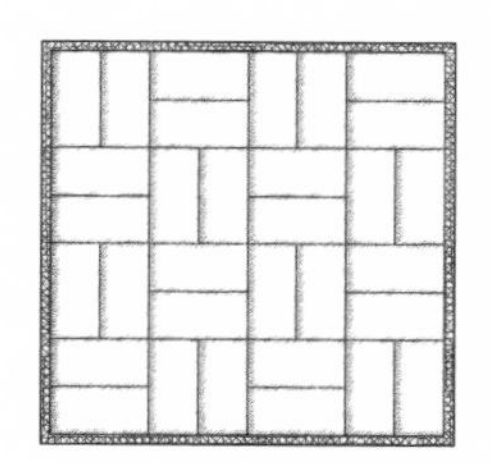

A chessboard can be tiled with dominoes each covering two squares. However, the indented board cannot.

Proof: In any tiling, a domino covers one white and one black square. Hence a tiling covers the same number of white and black squares. Since the indented board has two more white squares than black squares, it cannot be tiled. Q.E.D.

Pythagoras's Theorem

a proof by dissection

The Theorem of Pythagoras (ca. 569 – 475 B.C.E.) states that in a right-angled triangle the square on the hypotenuse, or long side, is equal to the sum of the squares on the other two sides (*opposite, top*). Nowadays this is written algebraically as $a^2 + b^2 = c^2$.

Proof: Arrange four identical right-angled triangles with sides a, b, and c in a large square of side $a + b$, leaving two square spaces with sides a and b, respectively (*opposite, center left*). The four triangles can also be arranged in this large square to leave a central square space with side c (*opposite, center right*). In both cases the contained squares equal the large square minus four times the triangle. Therefore the sum of the two smaller squares, $a^2 + b^2$, equals the larger square, c^2. Q.E.D.

Conversely, and this requires an extra proof, IF a triangle's sides are related as above, THEN it is right-angled. Integers that satisfy the equality $a^2 + b^2 = c^2$ are known as Pythagorean triples. An ancient construction of a right angle from a loop of string with $3 + 4 + 5 = 12$ equally spaced knots is based on the triple 3:4:5 (*below, left*). A Babylonian clay tablet, Plimpton 322, lists integer pairs corresponding to Pythagorean triples (*below, right*), which suggests that our general theorem may well have been known long before Pythagoras.

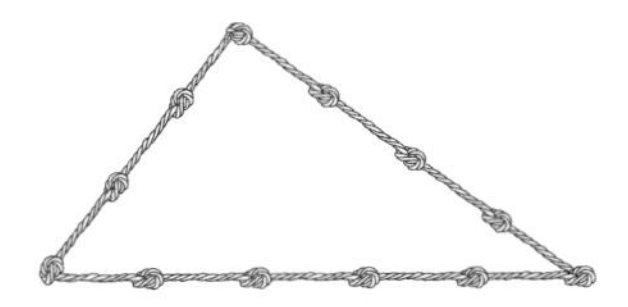

$$65^2 + 72^2 = 97^2$$

$$119^2 + 120^2 = 169^2$$

$$319^2 + 360^2 = 481^2$$

$$2291^2 + 2700^2 = 3541^2$$

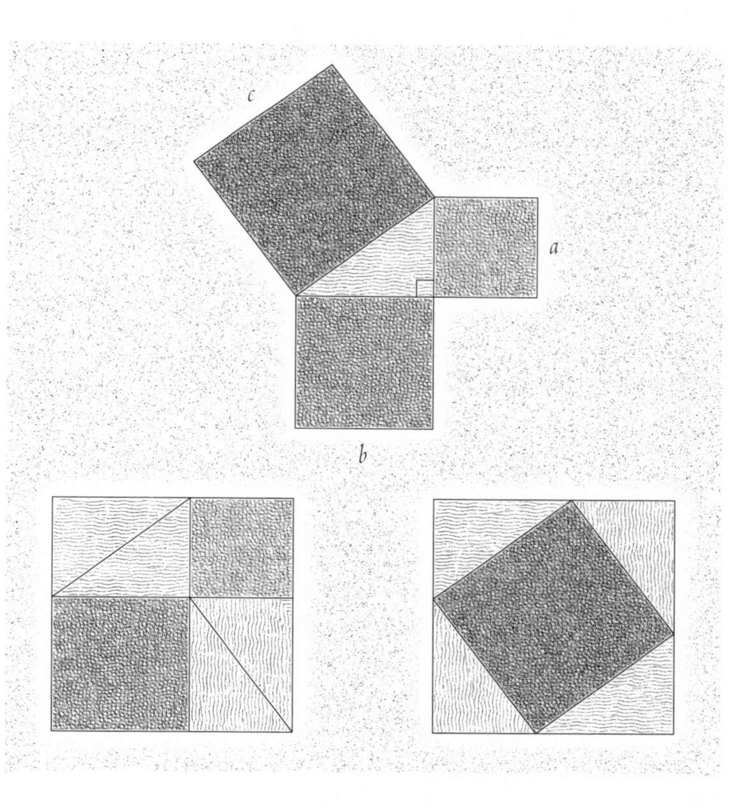

If, instead of squares, we fit any three similar figures to the sides of a right-angled triangle, we can also prove that the areas of the smaller two add up to that of the largest.

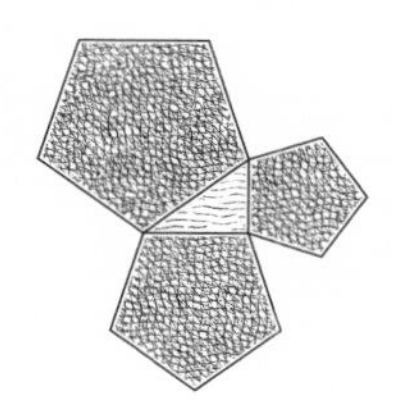

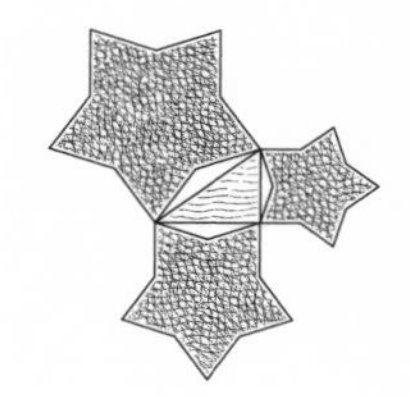

Plane and Simple

your basic theorem toolbox

The *Elements* of Euclid (ca. 325–265 B.C.E.) long ago set the standard for mathematical rigor, and, being a popular textbook ever since, much of its contents has been absorbed into our common cultural heritage.

Over the course of thirteen books, Euclid built up a complex network of theorems of ever-increasing depth, connected by logical arguments, and rooted in some intuitive facts, called *axioms* or *postulates*. To be prepared for the rest of this book, start with the four simple results on page 7, right, and, following the arrows, deduce the theorems on the left.

You also need to be able to recognize the two main levels of sameness of triangles. Two triangles are *similar* if they have equal angles. Since two angles in a triangle determine the third, you know that two triangles are similar if you can show that they share two angles. Two triangles are *congruent* if they have equal sides. This is the case whenever one of the five configurations of three sides and angles drawn solid below are present in both triangles. For example, the two gray triangles in the diagram (*below, right*) share one such configuration consisting of the two sides *r* and *m* and one right angle and are therefore congruent. Hence the two tangent segments *s* and *t* to the circle from the point outside have the same length.

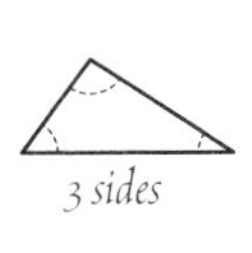

3 sides

2 sides and 1 angle

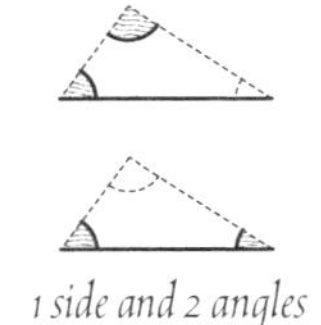

1 side and 2 angles

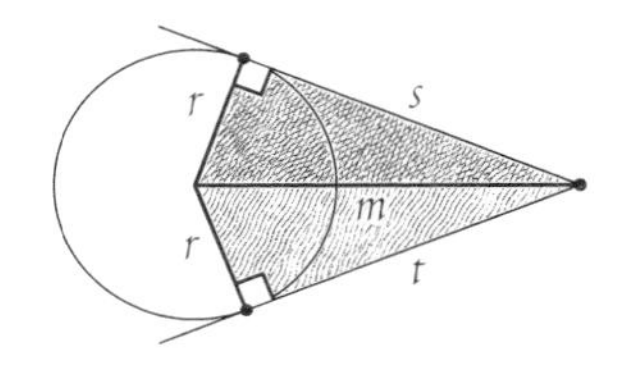

The Sum of the Angles in a Triangle

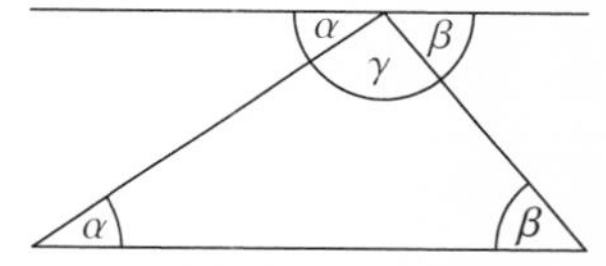

The sum of the angles is $\alpha+\beta+\gamma = 180^{0}$

Thales's Theorem

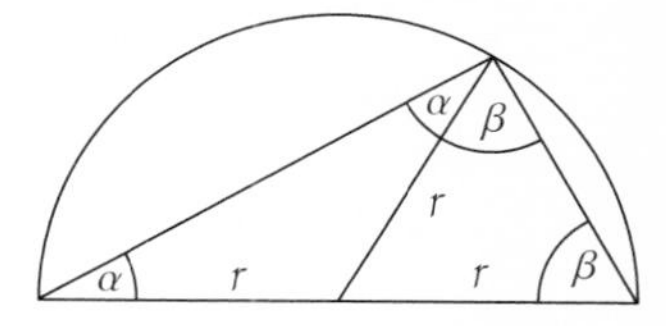

The upper angle $\alpha+\beta$ is 90^{0}

Squaring a Rectangle

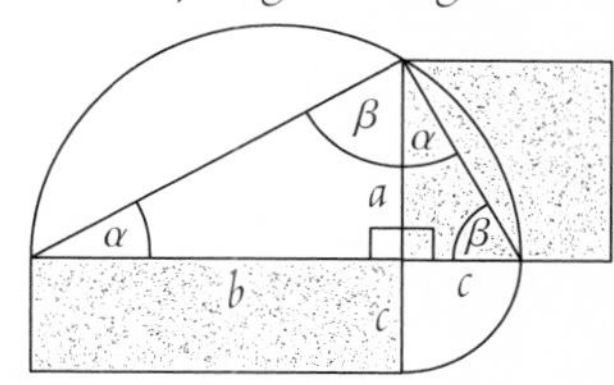

area square = $a^2 = b \cdot c$ = area rectangle
(by similarity of triangles $a/c = b/a$)

If the lines k and l are parallel

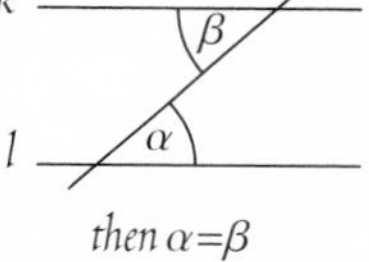

then $\alpha=\beta$

If

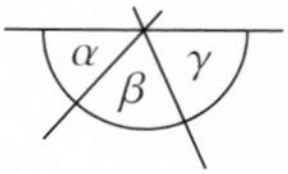

then $\alpha+\beta+\gamma=180^{0}$

If $a = b$

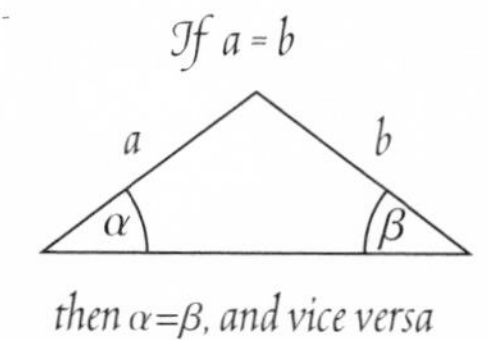

then $\alpha=\beta$, and vice versa

In similar triangles

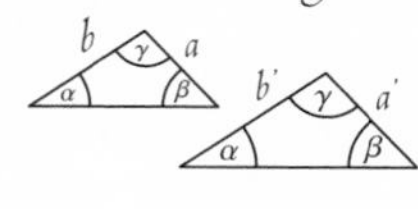

$a/a' = b/b'$

FROM PIE TO PI

mysteries of the circle

Eratosthenes (276 – 194 B.C.E.) is famous for his pizza pie method for calculating the circumference of the Earth based on the distance from Alexandria to Syene and the shadow angle at Alexandria at a time when the sun shone down a deep well in Syene, throwing no shadow there. Using the formula *circle diameter* $\cdot\ \pi =$ *circle circumference*, he also calculated the Earth's diameter. Fortunately, his pen pal Archimedes (287–212 B.C.E.) was able to prove a good estimate for the elusive value of π.

Since π is the circumference of a circle of diameter 1, it will be greater than the circumference of any inscribed and less than that of any circumscribed regular polygon (*opposite, top*). The more sides such a polygon has, the closer its circumference will be to that of the circle. Luckily, it is easy to calculate from the circumference of one such polygon the circumference of a polygon of the same kind with double the number of sides (*opposite, center*). Starting with regular hexagons, Archimedes successively calculated the circumferences of regular 12-, 24-, 48- and 96-gons to capture π between $3\frac{10}{71}$ and $3\frac{10}{70}$. The last of these values equals $\frac{22}{7}$ and is used even today in many school books instead of the true value of π. Using squares instead of hexagons, a formula for approximating π emerges (*opposite, bottom*).

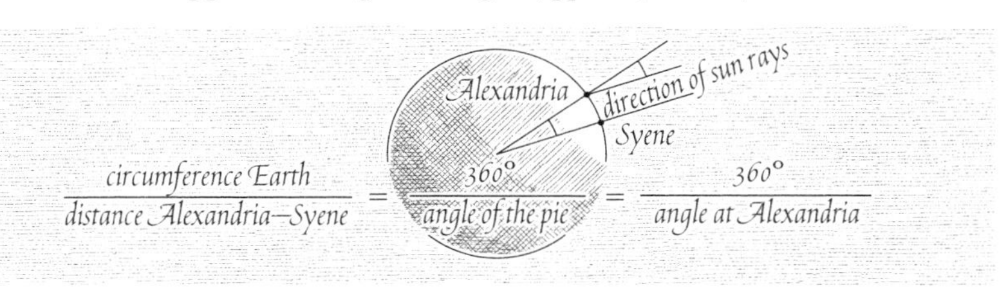

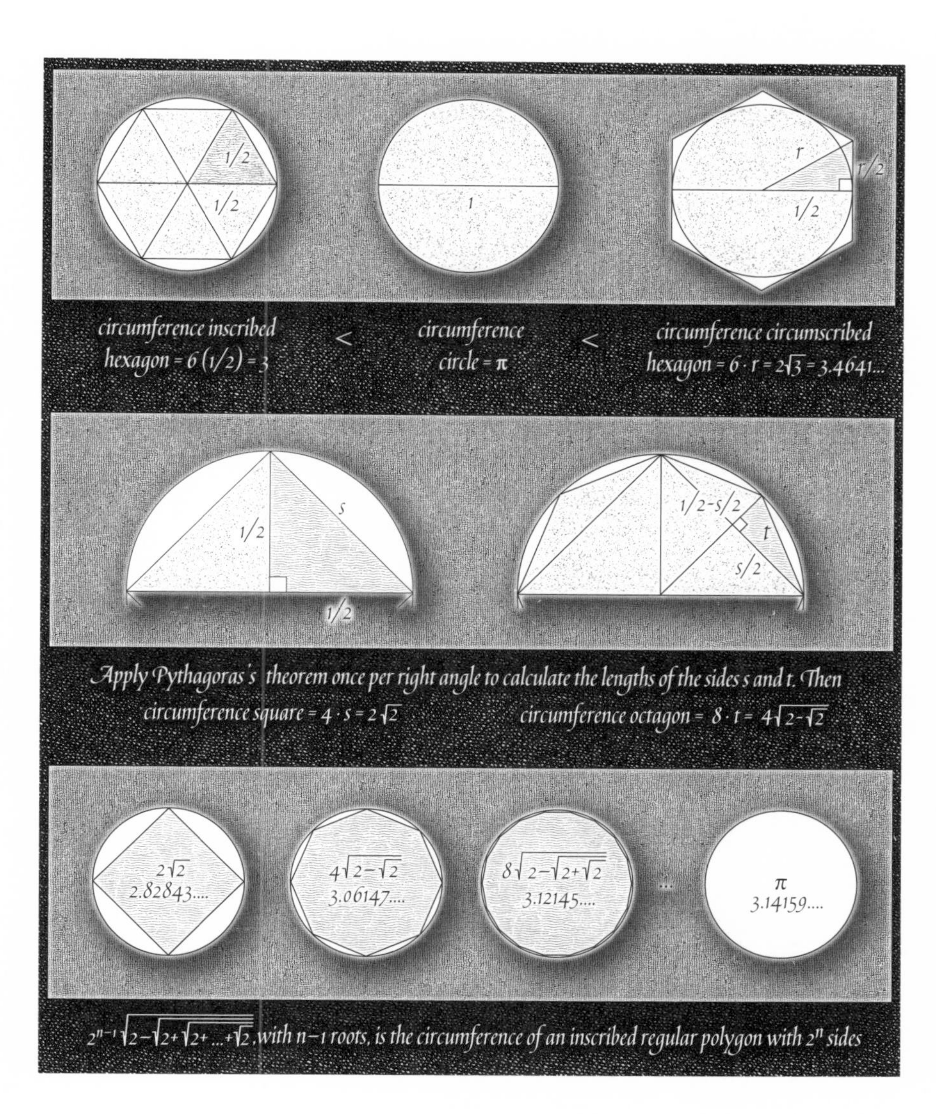
1/2
1/2
1
r
r/2
1/2
circumference inscribed hexagon = 6 (1/2) = 3
<
circumference circle = π
<
circumference circumscribed hexagon = 6 · r = 2√3 = 3.4641...
1/2
s
1/2
1/2−s/2
t
s/2
Apply Pythagoras's theorem once per right angle to calculate the lengths of the sides s and t. Then
circumference square = 4 · s = 2√2
circumference octagon = 8 · t = 4√(2−√2)
2√2
2.82843....
4√(2−√2)
3.06147....
8√(2−√(2+√2))
3.12145....
...
π
3.14159....
2ⁿ⁻¹√(2−√(2+√(2+...+√2))), with n−1 roots, is the circumference of an inscribed regular polygon with 2ⁿ sides

Cavalieri's Principle

a proof by approximation in slices

There are two versions of the celebrated principle named after Bonaventura Cavalieri (1598–1647). For plane figures it says that IF every horizontal line intersects two such figures in cuts of equal length, THEN the two figures will have equal area. Similarly, IF every horizontal plane intersects two solids in cuts of equal area, THEN the two solids will have equal volume.

An outline of the proof by approximation in slices, which is the same for both principles, is given on the opposite page. Cavalieri's principle is a good example of "divide (into manageable pieces) and conquer" in mathematics. For example, in the plane version, we reduce the difficult problem of calculating areas to the easier problem of measuring the lengths of line segments.

Below are some important area and volume formulas easily derived using Cavalieri's principle.

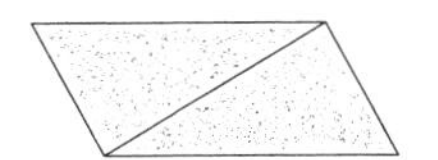
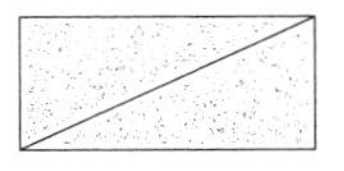

area parallelogram = area rectangle with the same base and height = base · height
area triangle = 1/2 area parallelogram = 1/2 base · height

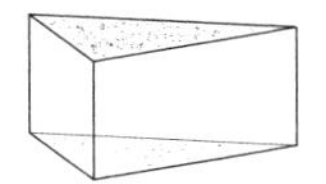
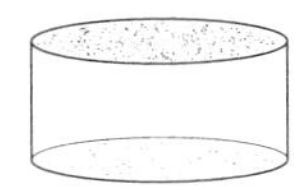
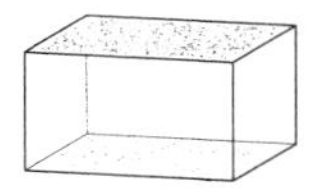

volume prism or cylinder = volume box with the same base area and height = base area · height

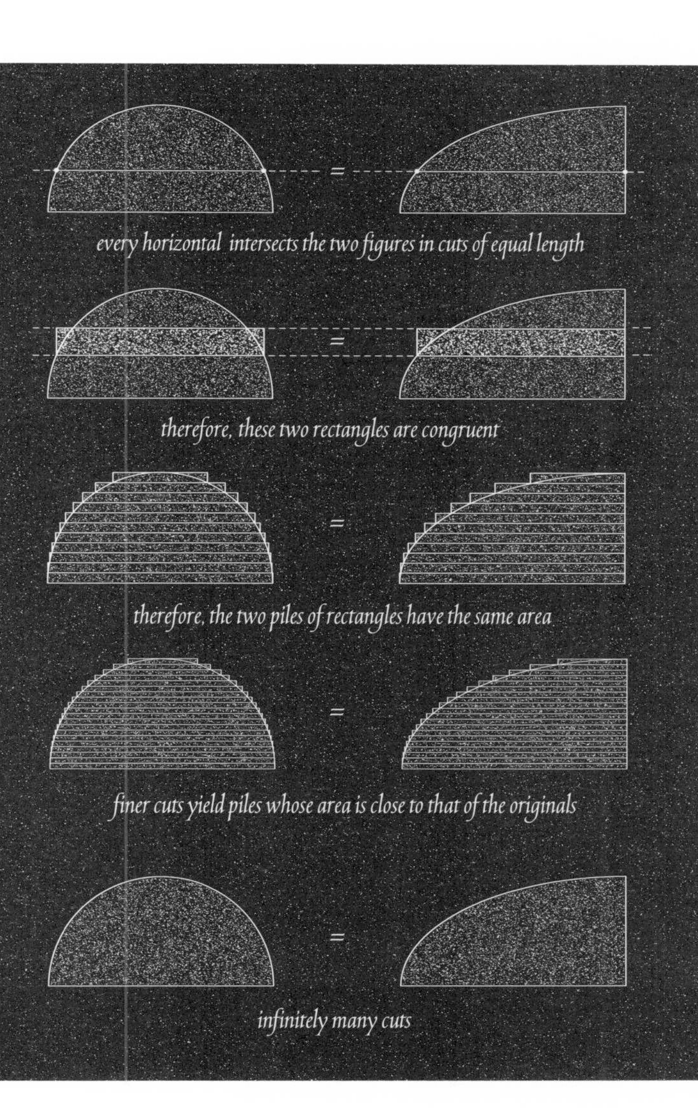
=
every horizontal intersects the two figures in cuts of equal length
=
therefore, these two rectangles are congruent
=
therefore, the two piles of rectangles have the same area
=
finer cuts yield piles whose area is close to that of the originals
=
infinitely many cuts

Cavalier Cone Carving

serious dissection in action

Cones come in all sorts of shapes and sizes: Piles of sand, limpet shells, pyramids, church spires, crystal tips and unicorn horns are all examples of cones. Every cone has a vertex and a base, which can be any plane figure.

Imagining the vertex as a beacon, a point is in the cone if its shadow is in the base. Let us prove that the formula for the volume of a cone is $\frac{1}{3}$ · *base area* · *height*, implying the volume formulas below.

A little shadow play (*opposite, top*) illustrates that all cones of the same height and base area are cut by any horizontal plane in slices of equal area. Now, Cavalieri's principle (*see page 10*) tells us that all these cones have the same volume. It is therefore enough to calculate the volume of one of these cones such as that of the right-angled pyramid (*opposite, center*). This and the other two pyramids combine into the triangular prism. Since all three pyramids have equal volume, this volume is one third of the volume of the prism. Q.E.D.

To cut a cube into six triangular pyramids of equal volume, first slice it with a diagonal plane into two triangular prisms and then slice those (as above) into three pyramids each. Or cut the cube into three identical square pyramids (*opposite, bottom*) and then slice each of those into a pyramid P_3 and its mirror image. These six pieces made from paper make a nice puzzle.

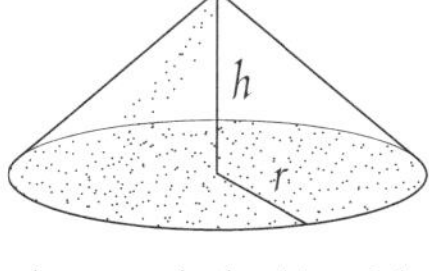

volume sand pile = $\frac{1}{3} \pi r^2 h$

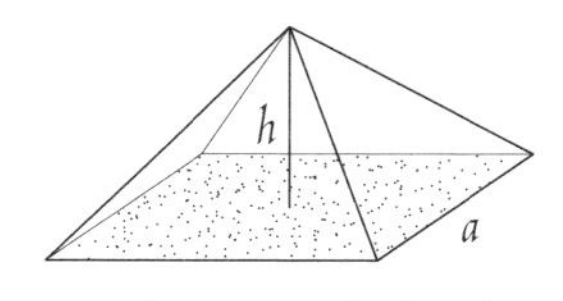

volume pyramid = $\frac{1}{3} a^2 h$

A punctual beacon projects figures of equal area in a plane to figures of equal area in any parallel plane; therefore the two cones have the same volume.

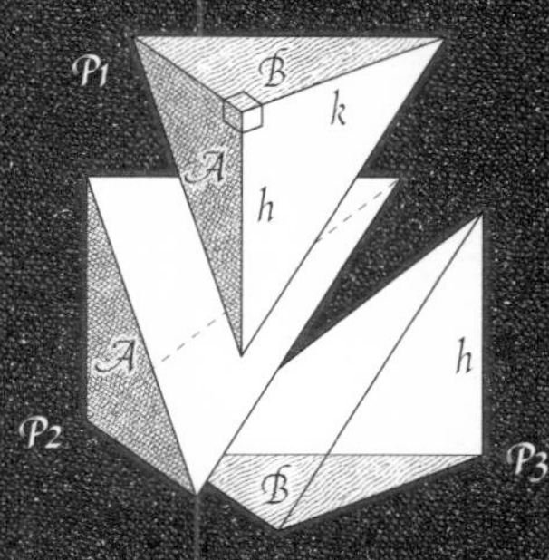

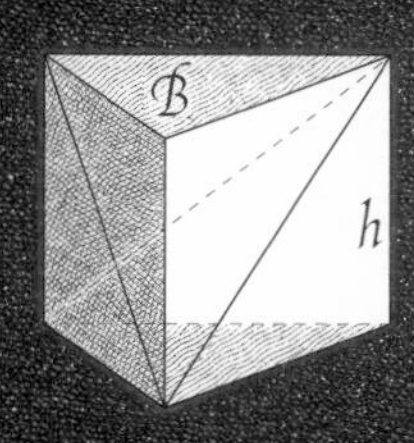

Pyramids P2 and P1 share the same base A and height k.
P1 and P3 also share base B and height h.
Volume of P1 = volume of P2 = volume P3 = 1/3 volume prism = 1/3 base area · height = 1/3 B · h.

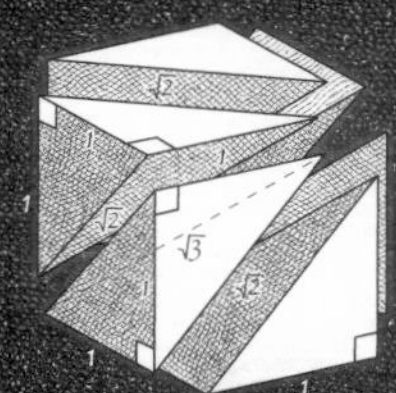

dissection of a cube into: three identical square pyramids (left), and: six triangular pyramids P3 (right)

A Frustrating Frustum

horses and moat walls

Many ancient manuscripts contain algorithms for calculating the areas or volumes of geometric figures, but not all formulas used by the ancients were correct. According to a Babylonian source, the volume of a *frustum*, or truncated square pyramid, was $(\frac{1}{2}(a+b))^2h$, whereas the Egyptian Rhind papyrus (ca. 1800 B.C.E.) indicates the descendants of the pyramid builders were using the correct version $\frac{1}{3}(a^2+ab+b^2)h$.

One of the oldest surviving Chinese mathematical treatises, *Jiuzhang Suanshu* 九章算術 (*Arithmetic in Nine Chapters*, ca. 50 B.C.E.), also mentions this version of the formula, and Liu Hui 劉徽 (ca. A.D. 263), in his commentary, gives a beautiful proof for it. He dissects the frustum or *fangting* 方亭 (square pavilion) into nine pieces: four identical pyramids or *yangma* 陽馬 (male horses), four prisms or *qiandu* 塹堵 (moat walls), and a rectangular box, all of which combine into a box and a pyramid. Adding the volumes of these together produces the volume of the frustum (*opposite, top*).

This proof assumes that we already know the formula for the volume of a square pyramid (*see page 12*), but we can nevertheless use our frustum dissection to rediscover this in an elegant snake-bites-its-own-tail way (*opposite, bottom*).

Other solids' volumes from Liu Hui's commentary are shown below.

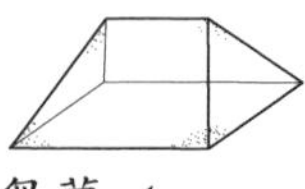

芻甍 *chumeng*
fodder loft

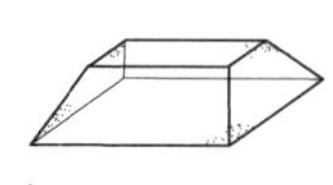

芻童 *chutong*
fodder boy

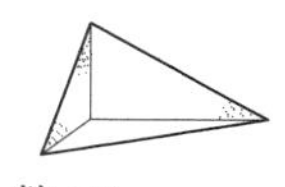

鱉腦 *bienao*
turtle's shoulder joint

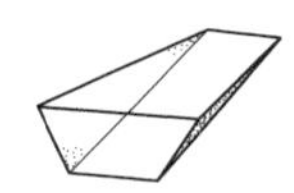

羨除 *xianchu*
drain

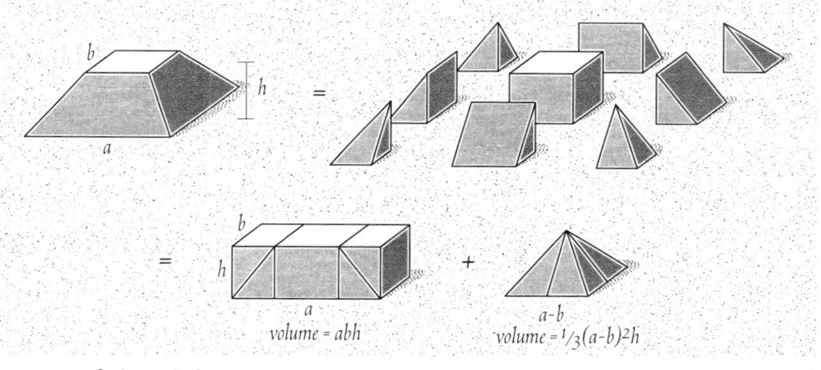

Liu Hui dissects the frustum into nine pieces: a rectangular box, four identical pyramids, and four prisms, all of which combine into a box and a pyramid, with respective volumes abh and $\frac{1}{3}(a-b)^2h$. Adding these together produces the volume of the frustum, $\frac{1}{3}(a^2+ab+b^2)h$.

Doubling the linear size of any plane or solid shape results in a fourfold increase in area, or an eightfold increase in volume. Using this we can slice a pyramid halfway as shown below.

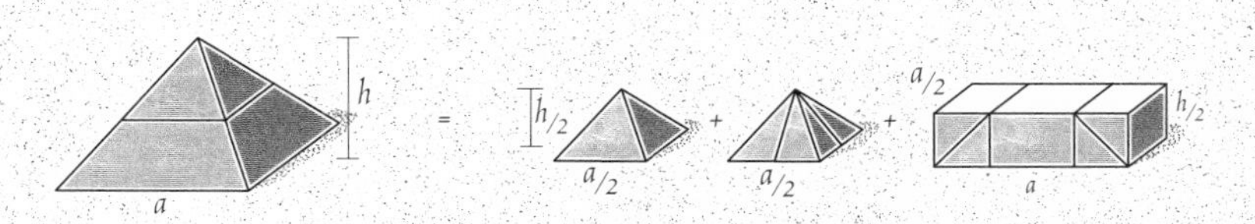

Volume of pyramid (v) = two $\frac{1}{8}$ volume pyramids + $\frac{a}{2} \cdot \frac{h}{2} \cdot a$. So, $\frac{3}{4}v = \frac{1}{4}a^2h$, which implies the volume of the pyramid, $v = \frac{1}{3}a^2h$.

ARCHIMEDES' THEOREM

mysteries of the sphere

Archimedes proved that the volume of a sphere is two-thirds that of the smallest cylinder containing it, and that its surface area is the same as that of the hollow cylinder. So taken was the philosopher with these relationships that he had a sphere and its surrounding cylinder inscribed on his tombstone.

Opposite we use Cavalieri's principle (*see page 10*) to derive the formula $\frac{4}{3}\pi r^3$ for the volume of a sphere of radius r, and thereby confirm Archimedes' first discovery.

For some real magic, project every point of the sphere, except the poles, onto another point on the cylinder, as shown below. Then it can be proved that any patch on the sphere gets projected onto a patch of equal area on the cylinder. If we then take the patch to be the whole sphere, it follows that its image is the cylinder, implying Archimedes' second discovery.

If we replace the sphere by a globe, project it onto the cylinder, and then slice the cylinder open, we find ourselves with a highly useful equal-area map of the Earth.

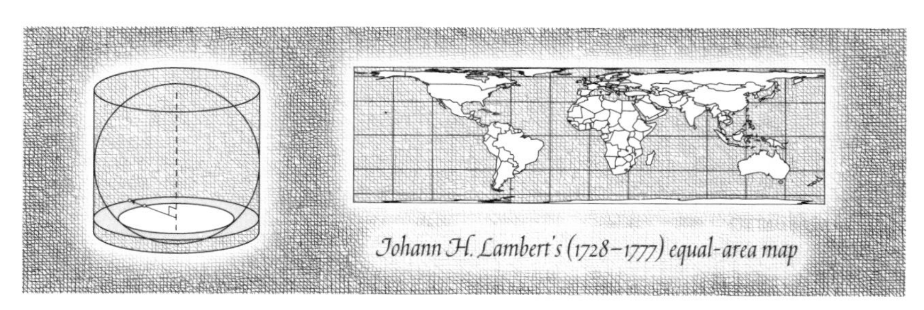

Johann H. Lambert's (1728–1777) equal-area map

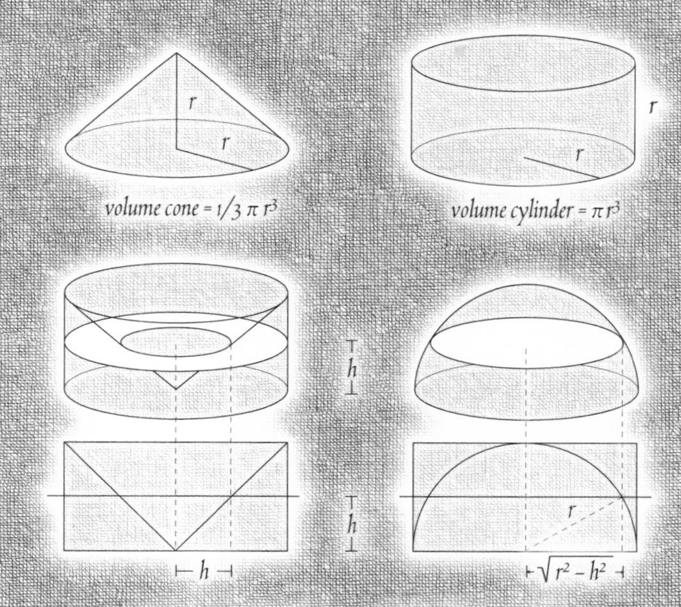

area ring $= \pi r^2 - \pi h^2 = \pi (r^2 - h^2) =$ *area circle*

Since the area of the ring is the same as that of the circle, Cavalieri's principle yields that the hemisphere and the cylinder minus the cone have the same volume.

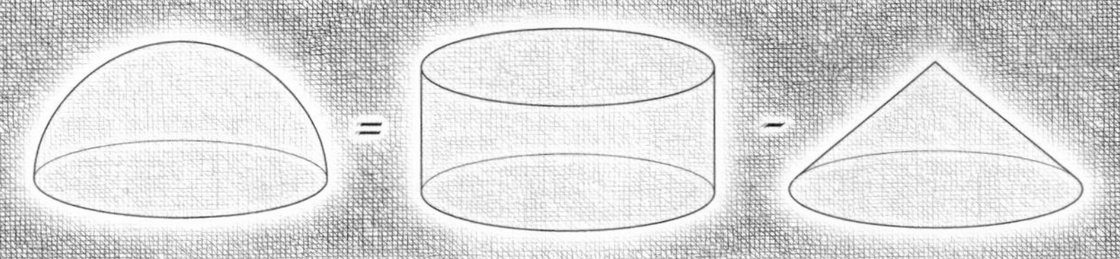

volume hemisphere = volume cylinder – volume cone = $\pi r^3 - 1/3\, \pi r^3 = 2/3\, \pi r^3$

volume sphere = 2 volume hemisphere = $4/3\, \pi r^3$

Inside Out

two proofs in wedges

Archimedes demonstrated how to mathematically kill two birds with one stone by using an ingenious idea to relate the insides and outsides of circles and spheres. Here is a sketch of his argument.

First he dissected the circle of radius r into a number of equal wedges (*opposite, top*) and arranged them into a roughly rectangular slab. He then observed that this can be done with an ever increasing number of wedges, and that as the number of wedges grows, the slab becomes indistinguishable from a rectangle whose short side has length r and whose long side is half the circumference of the circle. Therefore the area of the rectangle coincides with that of the circle, giving the formula:

$$\textit{area of circle} = \tfrac{1}{2} \cdot \textit{circumference of circle} \cdot r.$$

We arrive at the same result by calculating the area of the sawtooth figure—just note that one of the "triangles" has area $\tfrac{1}{2} \cdot \textit{base length} \cdot r$ and that the bases of the triangles sum to the circumference.

To derive a similar formula for a sphere of radius r, Archimedes dissected it into triangular cones whose common vertex is the center of the sphere and whose bases are contained in the surface of the sphere (*opposite, bottom*). These cones can play the role of the triangles in the sawtooth figure, and since (*from page 12*) the volume of one of these cones is $\tfrac{1}{3} \cdot \textit{base area} \cdot r$ we obtain the formula:

$$\textit{volume of sphere} = \tfrac{1}{3} \cdot \textit{surface area of sphere} \cdot r.$$

As the grand finale, we plug into our new relationships the formulas for the circumference of a circle of radius r and the volume of a sphere of radius r (*see page 16*) to conclude that the area of the circle is πr^2 and that the surface area of the sphere is $4\pi r^2$.

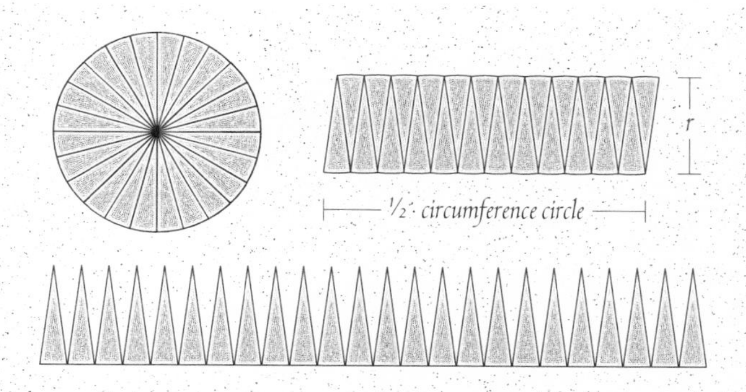

area of circle = area of wedges = ½ · circumference circle · r

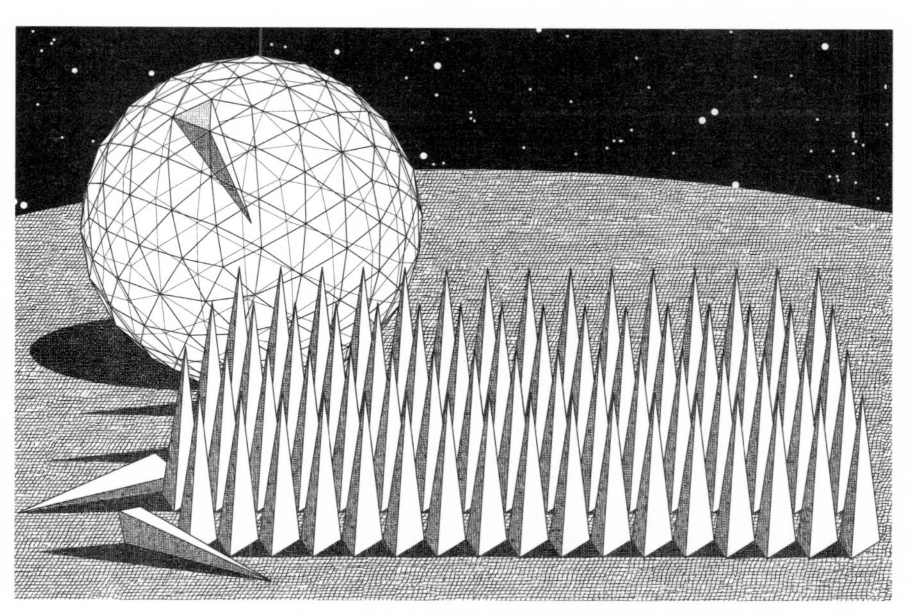

volume of sphere = volume of cones = ⅓ · surface area sphere · r

Mathematical Dominoes
proofs by induction

Setting up a number of dominoes in a row, one domino for each natural number, let's make sure that IF domino n topples, THEN so does domino $n+1$. If we now topple the first domino, we can be sure that every domino will eventually topple over.

Proof by induction is the mathematical counterpart of this insight. Now, instead of the dominoes, we have an infinite number of statements, one for each natural number. Here, we can be sure that all statements are true if we can prove that the first statement is true *and* the truth of statement n implies the truth of statement $n+1$.

The first three rows of diagrams (*opposite*) show how the first three statements corresponding to the following theorem imply each other:

THEOREM: Every 2^n by 2^n board that has been dented in one of its unit squares can be tiled with L-shapes made up of three unit squares.

PROOF BY INDUCTION: Since a dented 2 by 2 board is an L-shape (*opposite, top*) the theorem is true for $n=1$. Assuming that statement n is true and considering an arbitrary dented 2^{n+1} by 2^{n+1} board, we quarter it and remove three middle squares to create four dented 2^n by 2^n boards (*opposite, center*). By assumption these four boards can be tiled, and the four tilings extend to one of the 2^{n+1} by 2^{n+1} board. Q.E.D.

Some of the other tumbling patterns used in the art of domino toppling also translate into methods of proof. In the triangular pattern (*opposite, bottom*), for example, the front piece topples all other pieces. The corresponding method of proof can be used to show that Pascal's triangle, named after Blaise Pascal (1623 – 1662), is made up of binomial coefficients (*see also pages 42 and 54*).

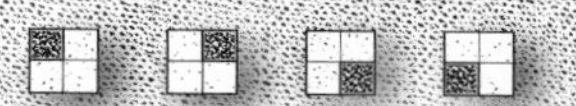

Every dented 2x2 board can be tiled with "L" shapes.

To show that an arbitrary dented 4x4 board can be tiled, quarter it and remove three middle squares to create four dented 2x2 boards and a tiling emerges. This suggests how to tackle the next case.

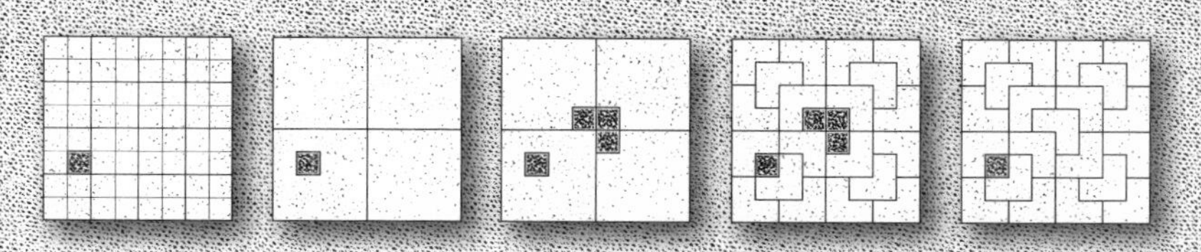

To show that an arbitrary dented 8x8 board can be tiled, quarter it and remove three middle squares to create four dented 4x4 boards. Extend tilings of these four to one of the 8x8 boards.

One proof that the number triangles are equal corresponds to tumbling the pattern on the left. Just note that in both number triangles any entry is the sum of the entries above and $1=\binom{0}{0}$.

The Infinite Staircase

a proof by regrouping

A classical paradox involves a number of identical bricks that are stacked up on top of a desk, as in the diagrams opposite. It is easy to prove that by adding more and more bricks as indicated, we can make the resulting staircase protrude as much as we want.

A staircase of n bricks, each of length 2, protrudes a distance of

$$1+\frac{1}{2}+\frac{1}{3}+\frac{1}{4}+\cdots+\frac{1}{n}$$

So, what we want to demonstrate is that the above sum approaches infinity as n does.

Proof: We first group the infinite sum as follows:

$$1+\frac{1}{2}+\left(\frac{1}{3}+\frac{1}{4}\right)+\left(\frac{1}{5}+\frac{1}{6}+\frac{1}{7}+\frac{1}{8}\right)+\left(\frac{1}{9}+\frac{1}{10}+\frac{1}{11}+\frac{1}{12}+\frac{1}{13}+\cdots\right.$$

We replace every term by a number less than or equal to it to produce a new sum that is less than or equal to the one we started with, and we notice that our substitutes add up to infinity:

$$1+\frac{1}{2}+\underbrace{\left(\frac{1}{4}+\frac{1}{4}\right)}+\underbrace{\left(\frac{1}{8}+\frac{1}{8}+\frac{1}{8}+\frac{1}{8}\right)}+\underbrace{\left(\frac{1}{16}+\frac{1}{16}+\frac{1}{16}+\frac{1}{16}+\frac{1}{16}+\cdots\right.}$$

$$1+\frac{1}{2}+\quad\frac{1}{2}\quad+\quad\frac{1}{2}\quad+\quad\frac{1}{2}\quad+\cdots$$

This means that the sum we are chasing is also infinite. Q.E.D.

Note that the staircase also gets infinitely tall as it grows infinitely broad and that actually building it gets very tricky very fast, involving closer and closer spacings of the bricks.

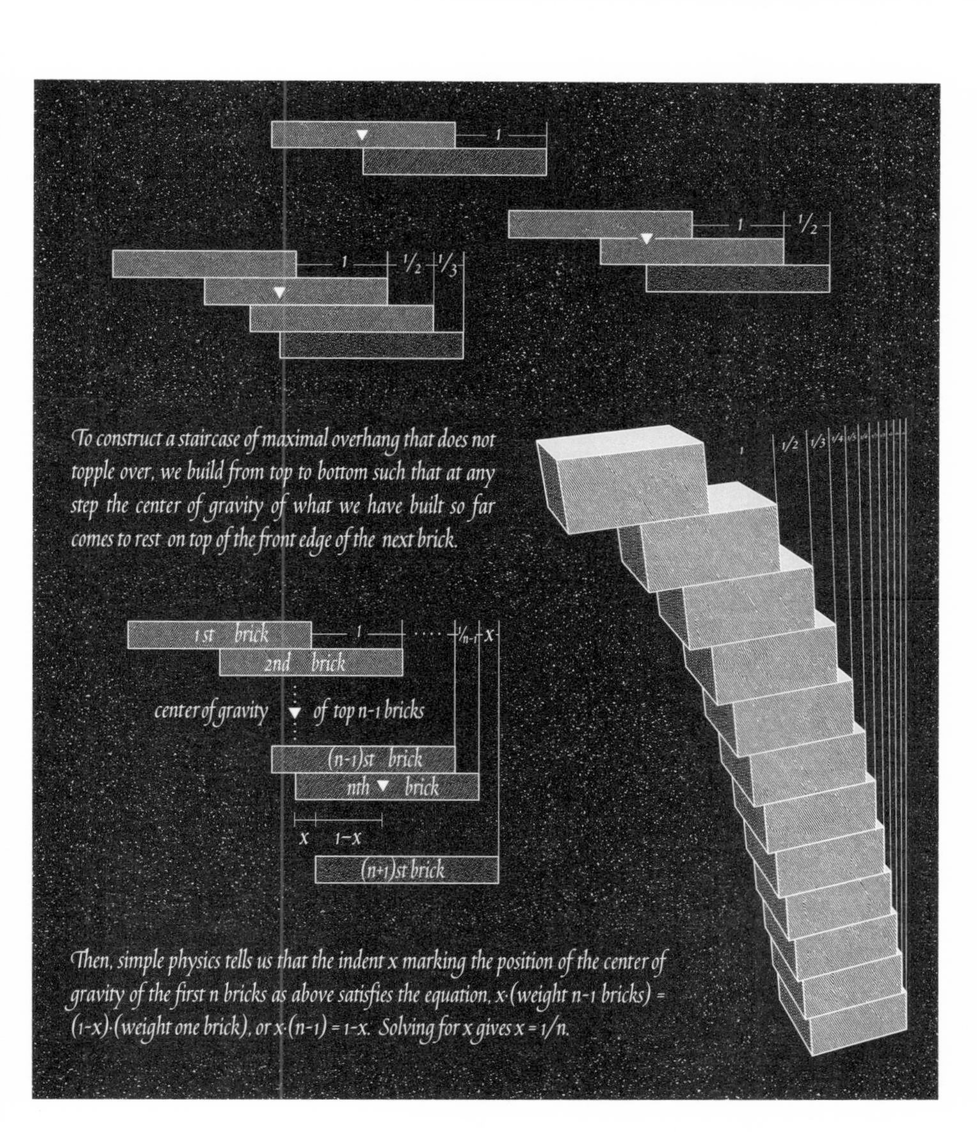

1
1
1/2
1
1/2
1/3
To construct a staircase of maximal overhang that does not topple over, we build from top to bottom such that at any step the center of gravity of what we have built so far comes to rest on top of the front edge of the next brick.
1/2
1/3
1st brick
1
1/n-1
x
2nd brick
center of gravity
of top n-1 bricks
(n-1)st brick
nth brick
x
1-x
(n+1)st brick
Then, simple physics tells us that the indent x marking the position of the center of gravity of the first n bricks as above satisfies the equation, x·(weight n-1 bricks) = (1-x)·(weight one brick), or x·(n-1) = 1-x. Solving for x gives x = 1/n.

Circling the Cycloid

a proof by dissection

Start with a regular polygon under a line, mark one of its top corners, and start rolling it along the line. Every time the polygon comes to rest on the line, indicate the position of the marked corner by a dot. Stop when the colored corner again touches the line and connect the dots by straight lines (*below, left*). Dissecting the polygon, it quickly becomes clear that the area enclosed by the resulting curve is exactly three times the area of the polygon (*opposite, top*).

Using a circle instead of a polygon, the resulting curve is a cycloid (*below, right*), used (with its relatives) by the ancient Greeks to describe the orbits of the planets. Since a circle can be approximated by regular polygons, the area enclosed by the cycloid is also three times the area of the circle.

The cycloid has many other important properties. For example, demonstrating the formidable power of Newton's and Leibniz's new infinitesimal calculus, Johann Bernoulli proved in 1696 that the cycloid is the solution to the difficult classical "problem of quickest descent." This means that if a particle slides along the cycloid from one of its ends to a second point, driven by gravity alone, it does so in less time than along any other curve connecting the two points.

Puzzle out the two one-glance proofs (*opposite, bottom*)!

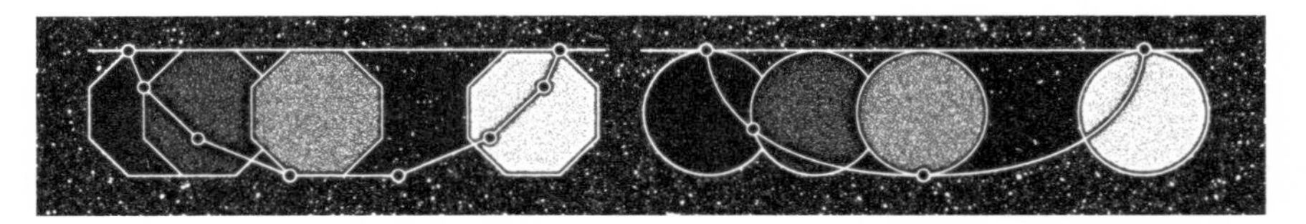

The three octagons fit under the curve produced by rolling one of them along the horizontal.

little square = 1/5 big square

area of regular 12-gon inscribed in circle of radius 1 is 3.

Slicing Cones

Dandelin's sphere trick

What kind of curve do you get when you slice a circular cone with a plane into an upper and a lower part? It may seem counterintuitive, but this shape will always be an ellipse, that is, the kind of curve you get when you pin the two ends of a piece of thread to a desk, pull the thread taut with a pen and draw a closed curve (*below*). In other words, an ellipse is the set of all those points in the plane the sum of whose distances from two fixed points (the focal points) is a constant.

To prove the slicing-cone theorem, Germinal Dandelin (1794–1847) inscribed two spheres into the cone that touch the slicing plane at one point each (*opposite, top*). He then observed that the cut is indeed an ellipse with these two points as focal points and associated constant the distance between the circles in which the two spheres touch the cone.

Similar tricks show that a plane cuts the cone in an ellipse, a parabola, a hyperbola (*opposite, bottom*) or, if it contains the vertex, a point, a line, or a pair of lines. Newton proved that two celestial bodies will always orbit each other on one of these conic sections, e.g., every planet orbits the Sun on an ellipse with the Sun at one of the focal points.

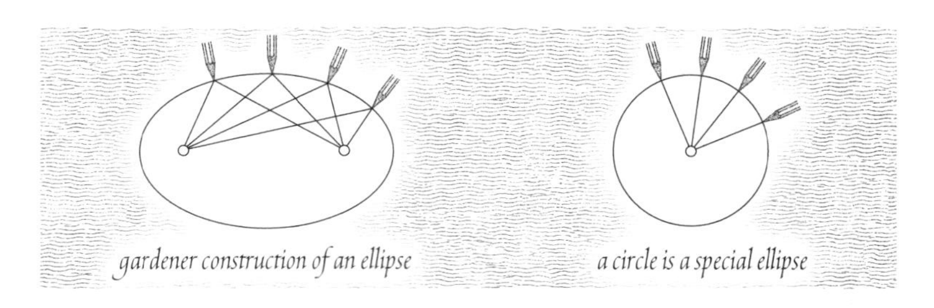

gardener construction of an ellipse — *a circle is a special ellipse*

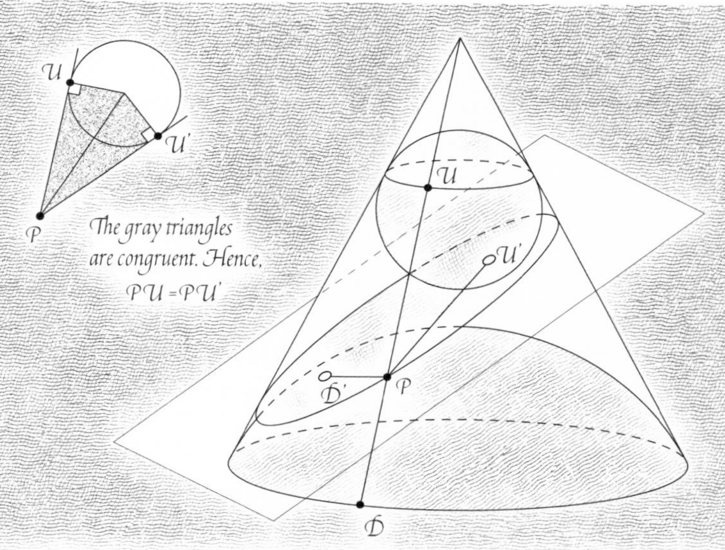

Since both the segments PU and PU' touch the upper sphere, PU' = PU. Similarly, PD' = PD. Therefore, PU'+PD'=PU+PD=UD, the distance between the touching circles of the spheres.

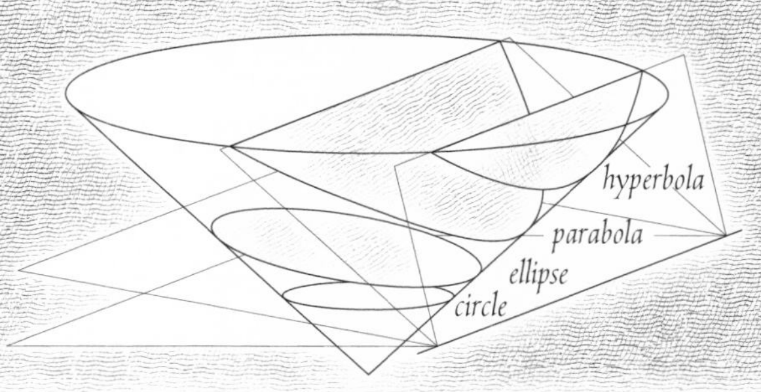

Tilting a plane up around the line, it first cuts the cone in a circle in the horizontal position, then ellipses, then a parabola in exactly one position, and from then on hyperbolas.

Folding Conics

burning mirrors and whispering walls

Mark a dot on a circular piece of paper, fold a perimeter point onto the dot to produce a crease line, and repeat for various points on the perimeter. An ellipse will begin to appear (*opposite, top*).

The proof that this works hinges on the defining property of ellipses in terms of two focal points (*see page 26*) and the property that motivates the term "focal point": If we make the inside of an ellipse into a mirror, every light ray originating at one of its two focal points will pass through the other, having been reflected off the ellipse (*below, left*). This is the principle behind burning mirrors and whispering walls. Put a candle at one of the focal points and its heat will focus at the other, whisper in one of the focal points of a large elliptical wall and your friend in the faraway other focal point will be able to hear you clearly. In general, light rays that miss the focal points envelope ellipses or hyperbolas sharing the focal points with the initial ellipse (*below, center and right*).

If you replace the circular piece of paper with a rectangular one and fold from one of its sides only (*opposite, bottom*), you get part of a parabola. From this, we can reconstruct the definition of a parabola in terms of a focal point and a line, and see how Archimedes came up with the idea of a parabolic array of mirrors focusing the sun to burn warships.

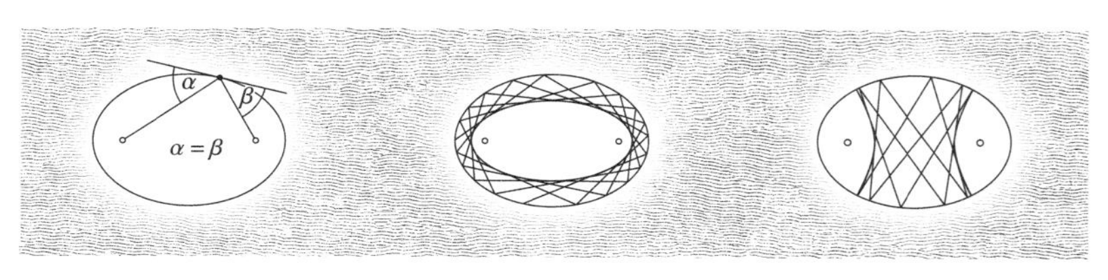

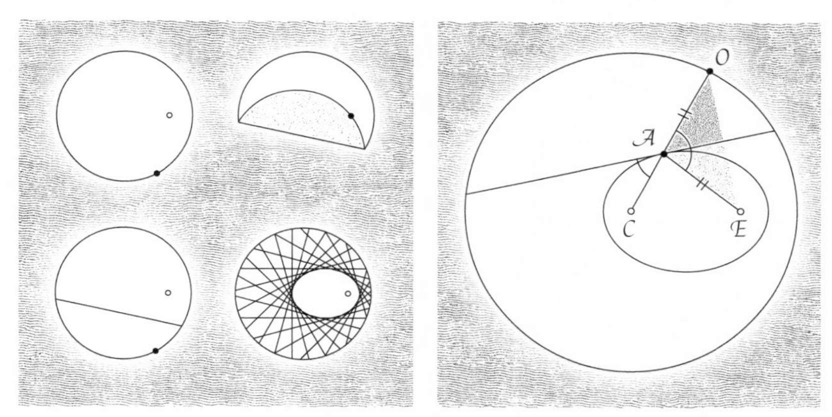

Since AO folds onto AE, AE + AC = AO + AC = CO = radius circle. Hence, as we move O around the circle, the point A describes an ellipse with focal points C and E and associated constant this radius. The three marked angles at A are equal. Therefore the creases are tangents of the ellipse and as such envelope it.

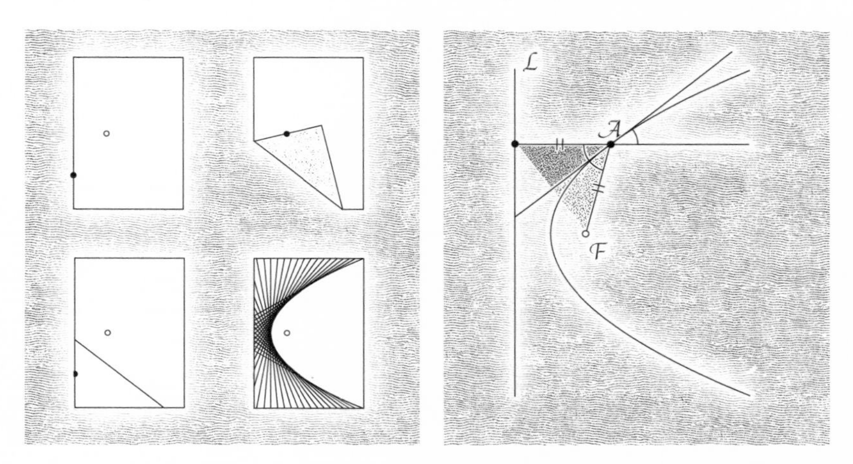

Folding of a parabola using its focal point F and base line L: (1) a point A of the parabola is at equal distance from F and L; (2) every horizontal light ray passes through F after being reflected off the parabola.

Knotting Polygons

a proof by paper folding

It is very easy to construct equilateral triangles, squares, and regular hexagons in a large number of different ways. Regular pentagons are more tricky, but here is the simplest way to construct one.

Tie a knot into a piece of tape and pull on the ends until the knot is completely flat. Cut off the excess tape on both sides and you are left with a regular pentagon! Why does this work?

Consider two regular pentagons sharing one side together with a piece of tape running through both of them (*below, left*). If we fold the left pentagon onto the right along the common side, the paper strip will neatly align itself along one of the sides of the right pentagon. Therefore, if we keep folding the tape around this pentagon, we will successively define all its sides and diagonals. Unwrapping the now creased tape and discarding the pentagons, we can finally tie the tape into a knot and flatten it such that no new creases appear.

The regular polygons with more than five sides can also be knotted using one or two paper strips, but the practical execution of these constructions can get very awkward.

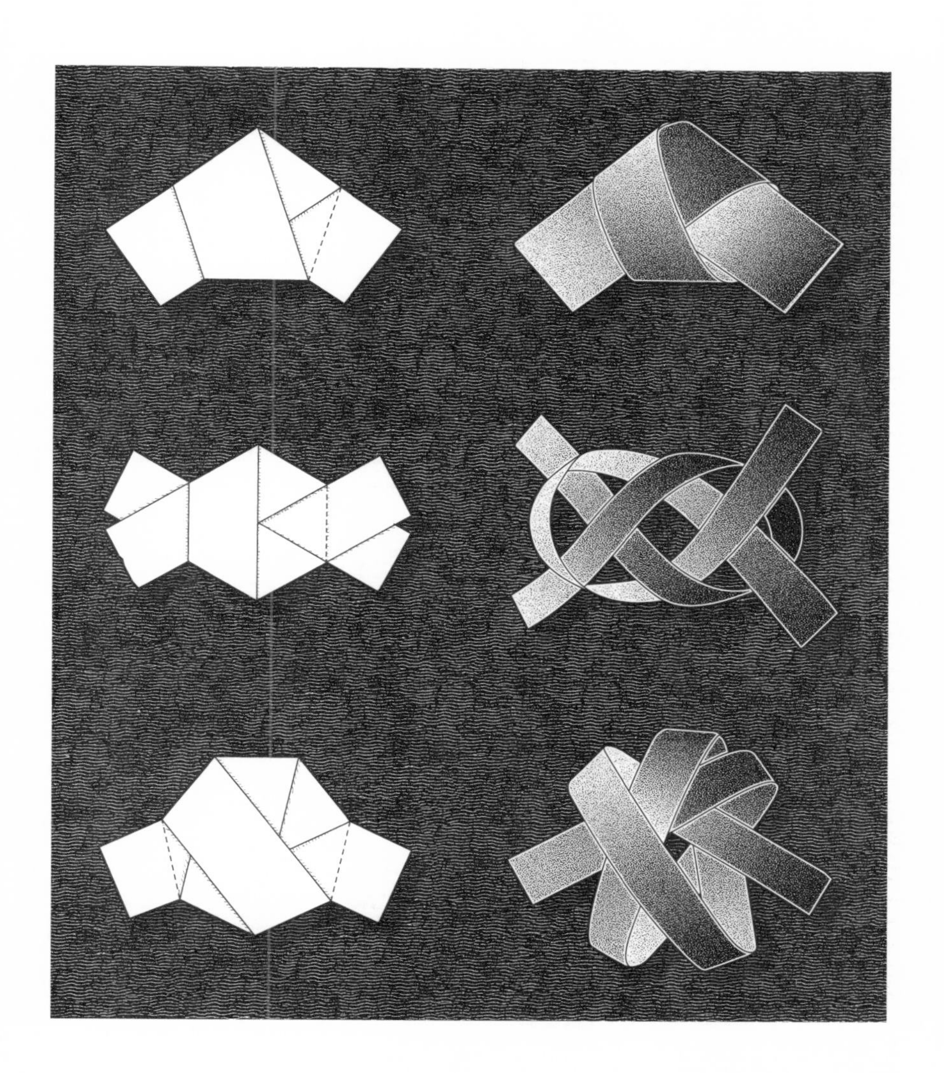

CUTTING SQUARES

a fresh look at an old pattern

Beautiful theorems are often arrived at by coming up with new ways of interpreting old patterns. For a whirlwind tour of some classic examples dating back to the Pythagoreans, let's consider various ways of dissecting a square array of n times n, or n^2, pebbles.

The first way gives the elementary equality $n+n+\ldots+n$ (n times) $= n^2$.

The second way translates into the surprising fact that the sum of the first n odd numbers is equal to n^2. For another proof of this theorem, just note that the numbers of triangles in the columns of tiling n below are the first n odd numbers and that after separating the black and the gray triangles (*opposite, bottom*), we get a square of n^2 triangles.

Closely related is the third way of dissecting, which corresponds to the equality $(n-1)^2 + (2n-1) = n^2$. Choosing the odd number $2n-1$ to be a square, we get a Pythagorean triple (*see page 4*). For example, choosing $2n-1 = 3^2$ gives $n = 5$, and therefore $4^2 + 3^2 = 5^2$.

One last way of dissecting the square array shows that n^2 is equal to the sum of the first n natural numbers plus the sum of the first $n-1$ natural numbers. Can you see how to derive from this a formula for the sum of the first n natural numbers?

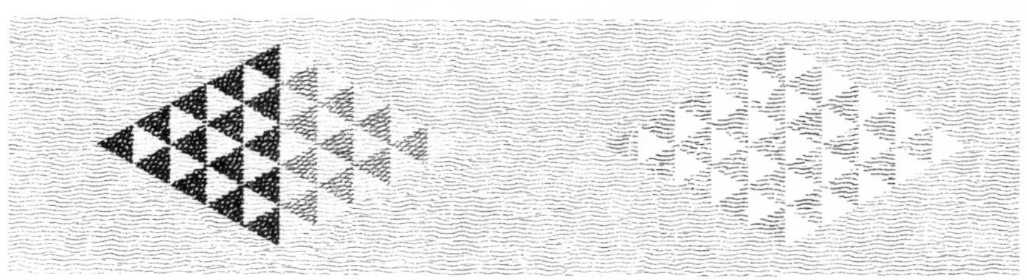

Power Sums
proofs by double counting

The marvelous Pythagorean one-glance dissection proof below shows that the sum of the first n natural numbers is half the number of pebbles in the rectangle, that is, $n(n+1)/2$. Carl Friedrich Gauss (1777–1855), one of the giants of mathematics, rediscovered this formula at the age of ten. Asked by his teacher to sum up the first 100 natural numbers, he made short work of the tedious task by observing that

$$1+100 = 2+99 = \ldots = 50+51 = 101,$$

and that therefore the required sum was $50 \cdot 101 = 5050$. This reasoning corresponds to looking carefully at the rectangle shown below, row by row ($1+4 = 2+3 = 5$ and the sum is $2 \cdot 5 = 10$).

The first diagram on page 35 elegantly shows that three times the sum of the first n squares equals the number of pebbles in the rectangle, that is, $(2n+1)(1+2+ \ldots + n)$.

The second diagram on page 35 demonstrates that the sum of the first n cubes is equal to the sum of the first n natural numbers squared.

For formulas for the sums of the first n fourth powers, fifth powers, and so on, see page 55.

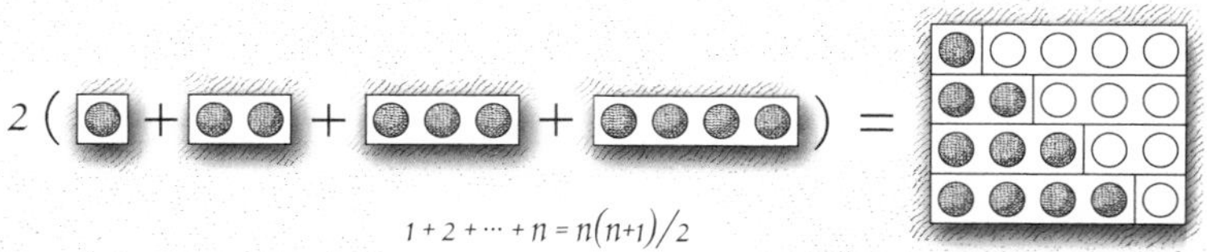

$1+2+\cdots+n = n(n+1)/2$

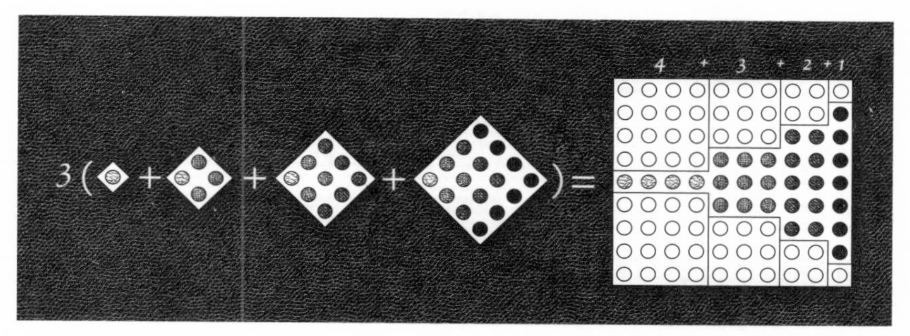

In general, $3(1^2 + 2^2 + \cdots + n^2) = (1 + 2 + \cdots + n)(2n + 1)$. *Plugging in* $1 + 2 + \cdots + n = n(n+1)/2$ *gives*

$$1^2 + 2^2 + \cdots + n^2 = n(n+1)(2n+1)/6.$$

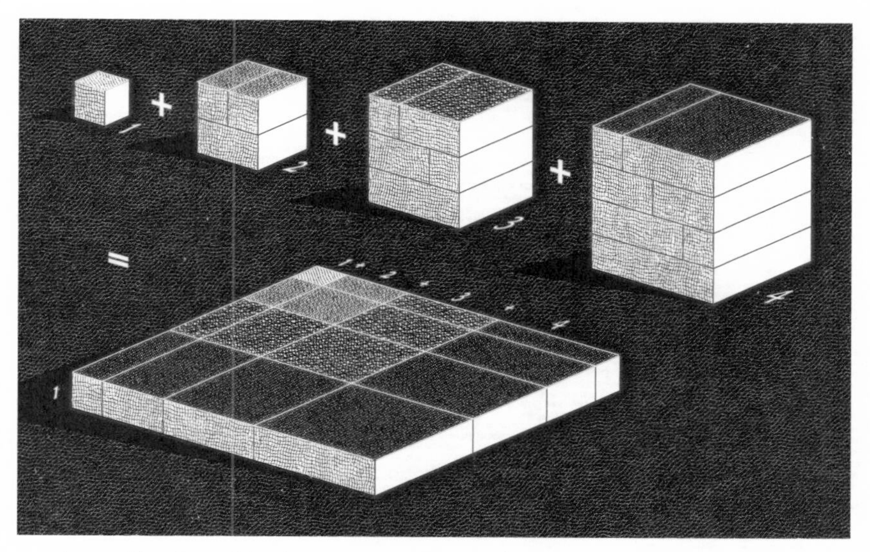

The sum of the volumes of the cubes, $1^3 + 2^3 + \cdots + n^3$, *equals* $1 \cdot (1 + 2 + \cdots + n)^2$, *the volume of the square slab.*

$$1^3 + 2^3 + \cdots + n^3 = (1 + 2 + \cdots + n)^2$$

Never-ending Primes

a proof by contradiction

Just as every object of the real world can be split in a unique way into indivisible atoms, every natural number can also be written in a unique way as the product of indivisibles called primes (the number 1 being an exception). The eight smallest primes are 2, 3, 5, 7, 11, 13, 17, and 19. The Sieve of Eratosthenes (*opposite*) is an elegant method for constructing all primes.

Euclid's *Elements* contains the following classic proof by contradiction that the world of numbers, unlike the real world, contains infinitely many primes.

Proof: There are either finitely or infinitely many primes. Assume that there are only finitely many and multiply all of them together to form a very large integer $n = 2 \cdot 3 \cdot 5 \cdot 7 \ldots$. Now, since $n+1$ is greater than any of the factors of n it cannot be prime, so one of the factors of n also has to be a factor of $n+1$. But, if this were so, then $(n+1) - n = 1$ would also have the same factor. This is a contradiction, so we conclude that our assumption of finitely many primes must be false. Hence there are infinitely many prime numbers. Q.E.D.

A *twin* of primes are two primes with a difference of two such as 5 : 7 and 11 : 13. Eternal fame awaits whoever can prove (or disprove) that there are infinitely many twins.

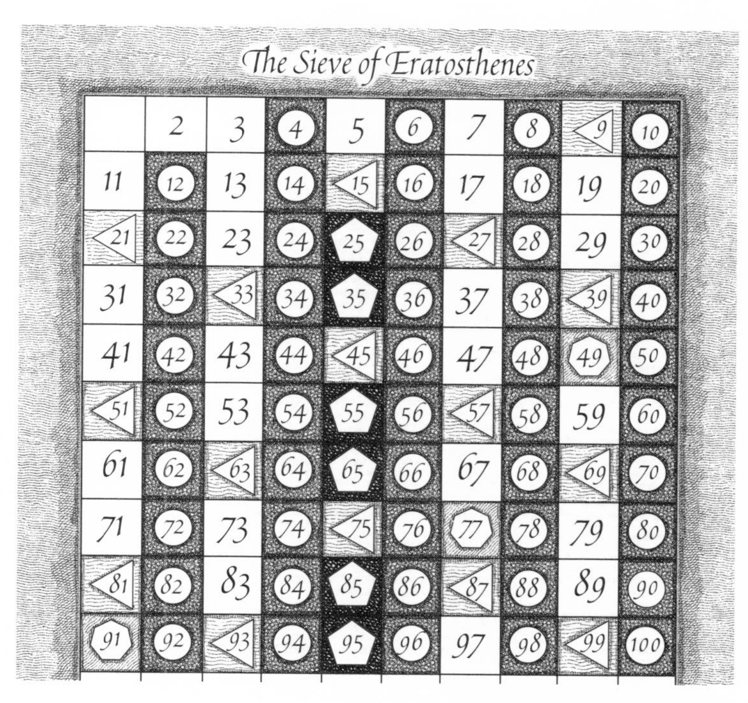

Circle all multiples of 2 in the above list. The smallest integer greater than 2 that is not circled is 3. Circle all its multiples. Now, the smallest integer greater than 3 that is not circled is 5. Circle its multiples, and so on. Then the primes are exactly the numbers that never get circled.

The Nature of Numbers

another proof by contradiction

On the number line (*below*), every point represents one of the real numbers that we use for measuring distances, areas, and volumes. By dividing the intervals between the integers into two parts, three parts, four parts, and so on, we single out the fractions, or rational numbers.

Even the tiniest patch of the number line contains infinitely many rational numbers and it may therefore seem reasonable to expect that every real number is rational. The Pythagoreans reputedly sacrificed a hecatomb, or one hundred oxen, to celebrate the discovery of a proof that √2, the length of the diagonal of a unit square, is irrational, so not a rational number.

Our proof (*opposite*) is an example of a proof by contradiction. We start by assuming that √2 is rational. This first implies the existence of an integer square (a square with integer diagonals and sides) and eventually a contradiction, that is, a statement that is not true. We conclude that our assumption is false. Therefore, √2 is irrational.

In general, it can also be shown that if a natural number is not a square, then its square root is an irrational number. This means that infinitely many of the radii of the root spiral (*opposite, bottom*) are irrational. Also, it turns out that, in a sense, there are many more irrational numbers than there are rational ones.

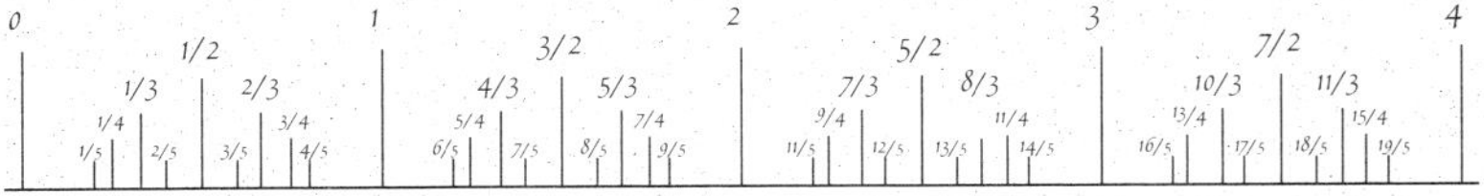

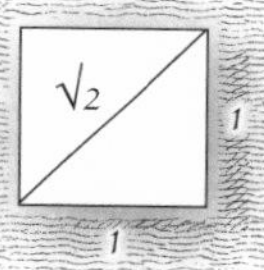

If $\sqrt{2}$ was a fraction b/a of positive integers, then the above square inflated by the factor a is the integer square below (left). By Pythagoras's theorem, $a^2 + a^2 = 2a^2 = b^2$. Hence $a^2 = (b/2)b$ is an integer.

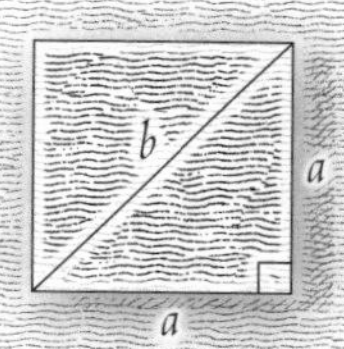

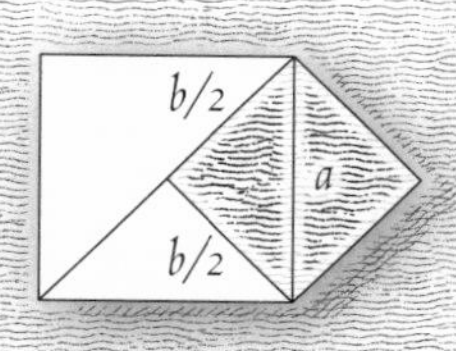

This is only possible if $b/2$ is an integer. Therefore, the small gray square on the right is also an integer square. Applying this construction to this second integer square gives a third, then a fourth, etc.

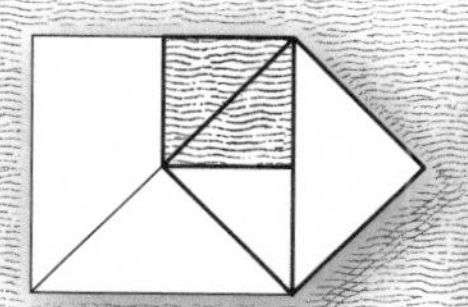
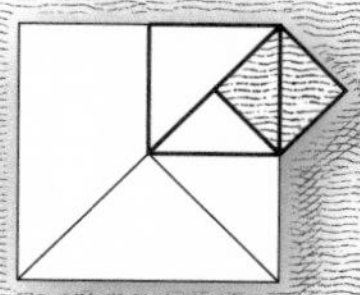
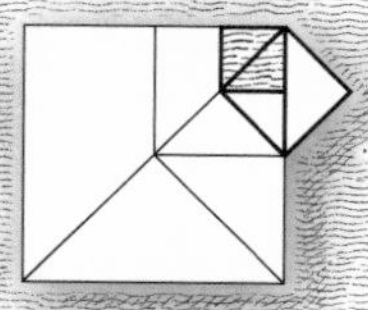
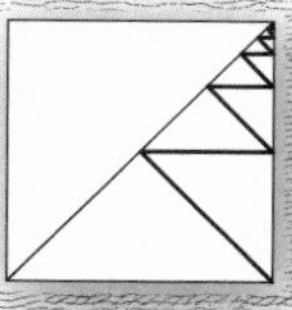

Every segment of the infinite zigzag on the right is a side of one of our integer squares and therefore has integer length. Impossible since the segments get infinitely small, whereas the smallest positive integer is 1.

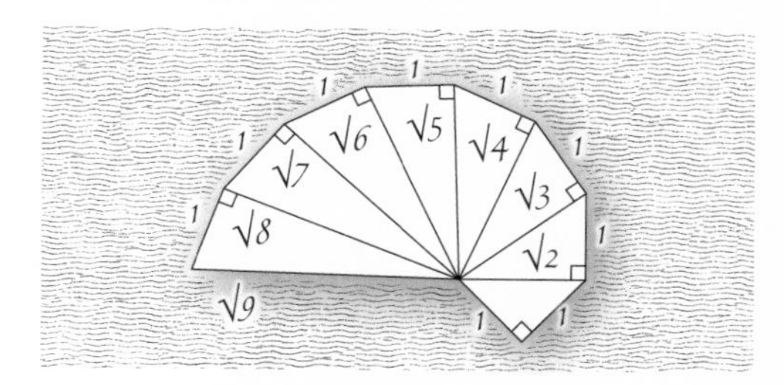

The Golden Ratio

nature's favorite number

What does a rectangle look like that is not too slim and not too wide, a rectangle that looks just right? For many artists and scientists this age-old beauty contest has a clear winner: the so-called golden rectangle (*below, left*) whose ratio of long to short side equals the golden ratio ϕ (*Phi*) of diagonal to side in a regular pentagon (*opposite, top*).

The golden ratio is present in many of nature's designs such as leaf arrangements and spiral galaxies. For example, if we take away a square from a golden rectangle (*below*), we find we are left with another golden rectangle since $\phi = 1/(\phi-1)$ (*opposite, top*). Repeating this process yields a spiral of squares that hugs many naturally occurring spirals.

Combining three golden rectangles at right angles (*opposite, center*), their twelve corners become the corners of an icosahedron. To prove this, we only have to check that all the triangles in the middle picture are equilateral, or equivalently, that the two essentially different edges of these triangles all have equal length.

This paves the way to a beautiful construction of an icosahedron from an octahedron (*opposite, bottom*) where the twelve corners of the former divide the twelve edges of the latter in the golden ratio.

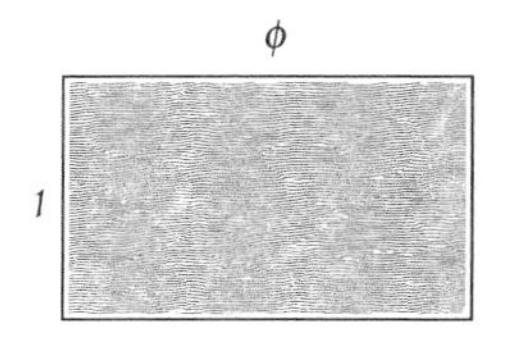

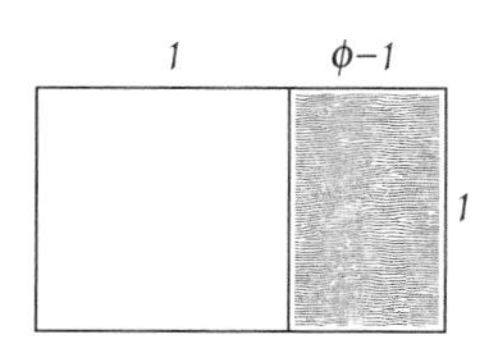

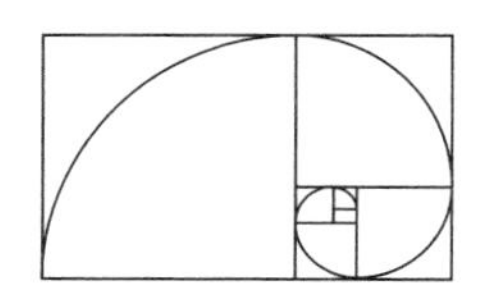

Angles α and β are equal. Therefore, the first two gray triangles are congruent and the second two similar. Hence $\phi/1 = 1/(\phi - 1)$, or $\phi^2 = \phi + 1$. Solving this equation gives $\phi = (1+\sqrt{5})/2 = 1.61803\ldots$

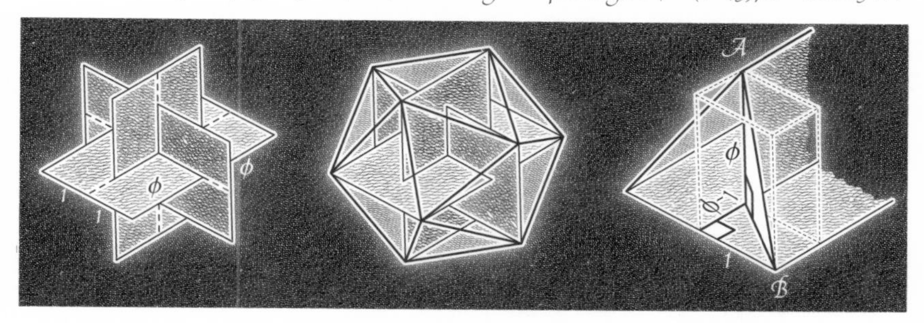

The highlighted edge AB has length $\sqrt{\phi^2+(\phi-1)^2+1^2} = \sqrt{2(\phi^2-\phi+1)}$ (apply Pythagoras's theorem twice) and since $\phi^2 = \phi + 1$, this is equal to 2, the length of the side of one of the three golden rectangles.

Put the three rectangles in squares as on the left. Then the edges of the squares form an octahedron.

The Numbers of Nature
the geometry of growth

A spiral of squares growing around a unit square as shown opposite left consists of squares the lengths of whose sides are the Fibonacci numbers 1, 1, 2, 3, 5, 8, 13, 21, . . . , named after Leonardo Fibonacci (1170–1250). Every number in the sequence is the sum of the two preceding it, so that $2 = 1+1$, $3 = 1+2$, $5 = 2+3$, and so on.

Fibonacci numbers are connected in many wonderful ways. For example, the tiling of the rectangle on the top left demonstrates that $1^2+1^2+2^2+3^2+5^2+8^2+13^2 = 13\cdot(13+8) = 13\cdot 21$. In general, the sum of the squares of the first n Fibonacci numbers equals the product of the nth and n+1st such numbers. Similarly, the tiling of the square on the right shows that $1\cdot 1+2\cdot 1+3\cdot 2+5\cdot 3+8\cdot 5+13\cdot 8+21\cdot 13 = 21^2$. This equality also generalizes easily.

The Fibonacci numbers often show up in the same phenomena as the golden ratio ϕ (*see page 40*) and it can be proved that the nth Fibonacci number is the closest natural number to $\phi^n/\sqrt{5}$. This implies that the rectangles we come across when building our spiral of squares become indistinguishable from golden rectangles.

Fibonacci numbers are hidden in many growth processes. For example, the numbers of clockwise and counterclockwise spirals apparent in sunflower heads (*opposite, center*) are usually consecutive Fibonacci numbers. Pascal's triangle (*opposite, bottom*) also grows, here row by row, with neighboring entries in one row adding up to the number below them. Since the sums of the first two diagonals of this triangle are both 1, and the sums of any two consecutive diagonals add up to the sum of the next, our golden sequence will appear yet again.

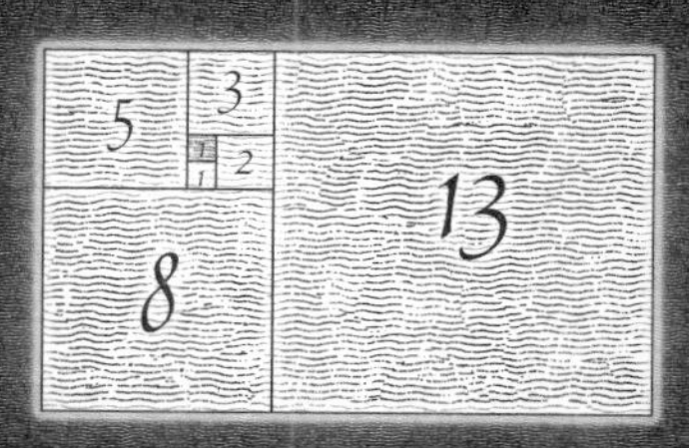

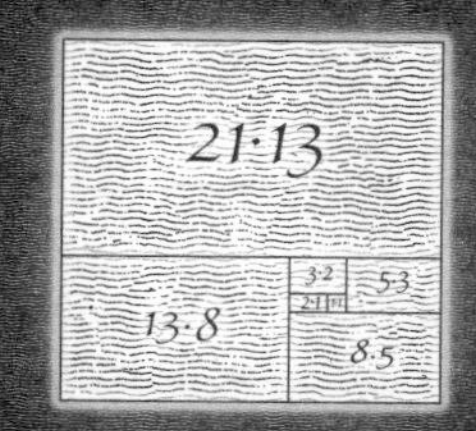

The different stages of the infinite spiral of squares form rectangles (left) the lengths of whose sides are two consecutive Fibonacci numbers. The first n such rectangles tile a square (right) if n is an odd number.

The numbers of clockwise and counterclockwise spirals visible in a sunflower head are (usually) consecutive Fibonacci numbers such as 34/21 on the left and 89/55 on the right.

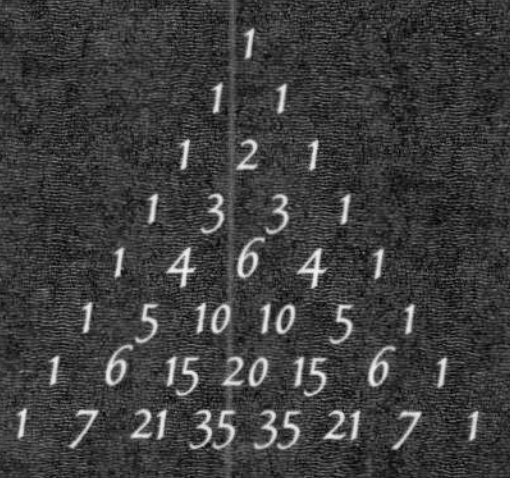

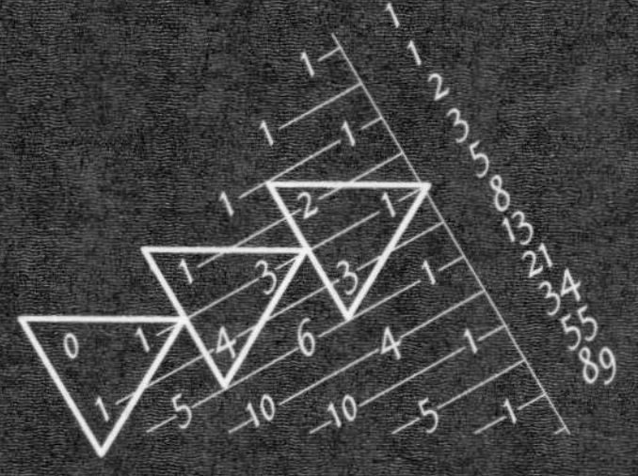

Every entry in a diagonal is the sum of one entry each from the previous two diagonals, as on the right. This shows that the sums of two consecutive diagonals add up to the sum of the following diagonal.

EULER'S FORMULA

a proof by pruning

A cut diamond is a solid without indentations all of whose faces are flat polygons. Leonhard Euler (1707 – 1783) discovered the neat formula that relates the numbers of vertices, edges, and faces of such a solid:

$$V(ertices) + F(aces) - E(dges) = 2$$

For example, in the case of a cube we count 8 vertices, 6 faces, and 12 edges and, indeed, $V+F-E = 8+6-12 = 2$.

PROOF: Start by opening out the network of vertices and edges to get a plane picture of the solid (*opposite, top*) in the form of a map with the same numbers of vertices, edges, and faces (the outside counts as one face). We notice that inserting a diagonal into a face yields a map with the same $V+F-E$ (*opposite, second row*) and so insert diagonals until a map consisting solely of triangles is produced. Finally, working around the outer border of the map and eliminating one triangle at a time (*opposite, lower two rows*) we are left with a map consisting of just one triangle (3 vertices, 2 faces, 3 edges). Since at every step the value of $V + F - E$ does not change, then $V+F-E = 3+2-3 = 2$. Q.E.D.

It is also not hard to show that Euler's formula holds true for any connected plane network of vertices and curve segments (*below*).

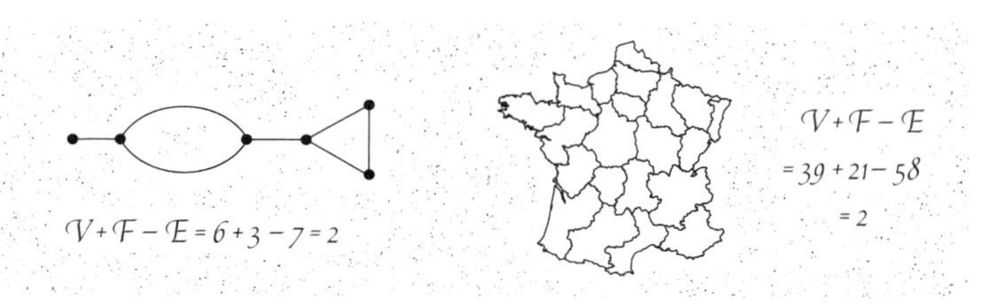

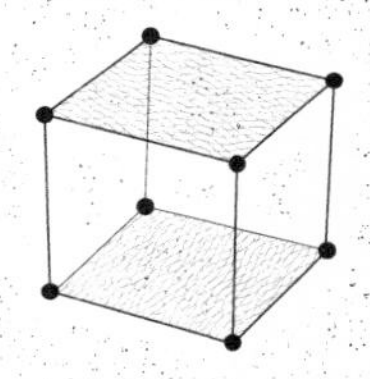

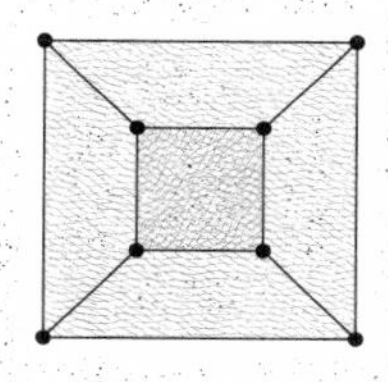

Opening out a cube gives a map with the same $V + F - E$.

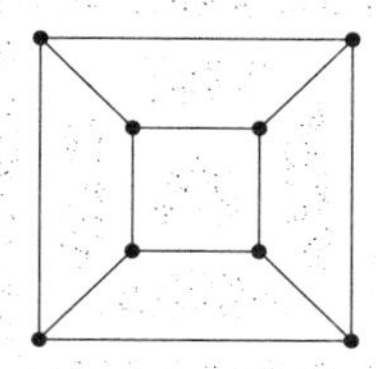

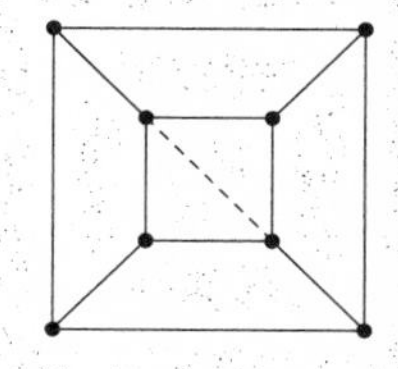

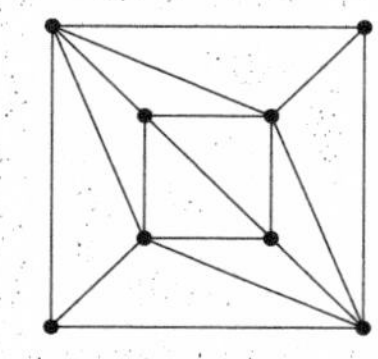

Inserting a diagonal increases both the number of faces and edges by one, leaving $V + F - E$ unchanged. Hence the $V + F - E$ of the triangulated map is the same as that of the original.

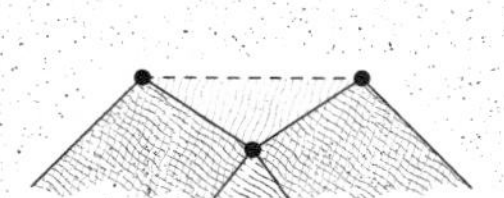

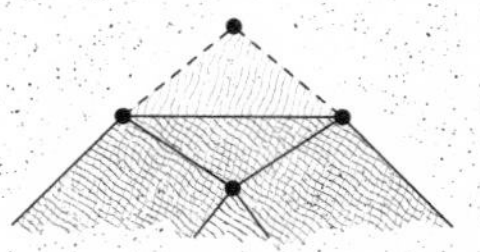

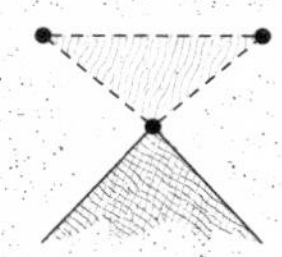

Pruning a triangle from the outside of a triangulated map leaves $V + F - E$ unchanged. For example, in the left diagram we lose one edge and one face: $V + F - E = V + (F - 1) - (E - 1)$.

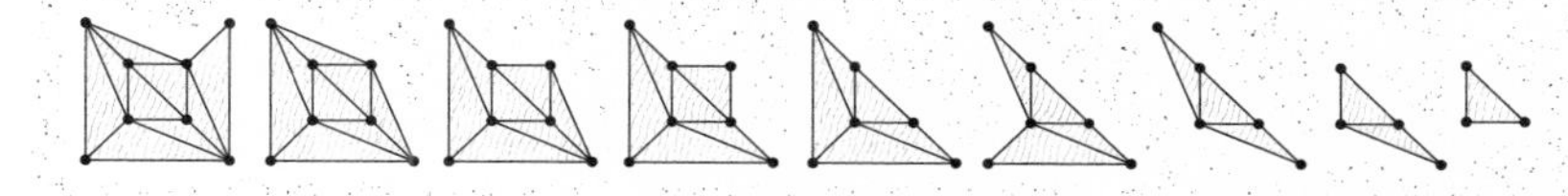

Serious pruning shows that $V + F - E$ of the original solid equals that of a single triangle.

POSSIBLE IMPOSSIBILITIES
doubling, squaring, and trisecting

Socrates (469–399 B.C.E.) once used the first two diagrams below to show how to double a square, and the oracle of Delphi predicted that a plague could be stopped by doubling the cubical altar of Apollo.

In the nineteenth century it was proved that "doubling a cube" as well as the other two notorious geometry problems of "squaring a circle" and "trisecting a general angle" are impossible if we require, like the ancient Greeks, that we use only a compass and unmarked ruler. If, however, we are allowed to use other tools, all three problems can be solved.

To square a circle, roll it half a revolution on a horizontal (*opposite, top*) to construct the long rectangle, which has the same area as the circle. Now, using compass and ruler as indicated, construct the square, which has exactly the same area as this rectangle (*see page 7*).

Archimedes discovered an ingenious method for trisecting an angle α between two intersecting lines (*opposite, center*) using a compass and a ruler with two marks on it: Just draw a circle and align the ruler as in the diagram. Then the angle ε is exactly one third of the angle α.

Doubling a square amounts to constructing $\sqrt{2}$ from 1 (*below, right*), and doubling a cube involves constructing $\sqrt[3]{2}$ from 1 (*opposite, bottom*). Just trisect and fold the paper unit square, as indicated, to construct $\sqrt[3]{2}$. Easy to describe but tricky to prove. Can you do it?

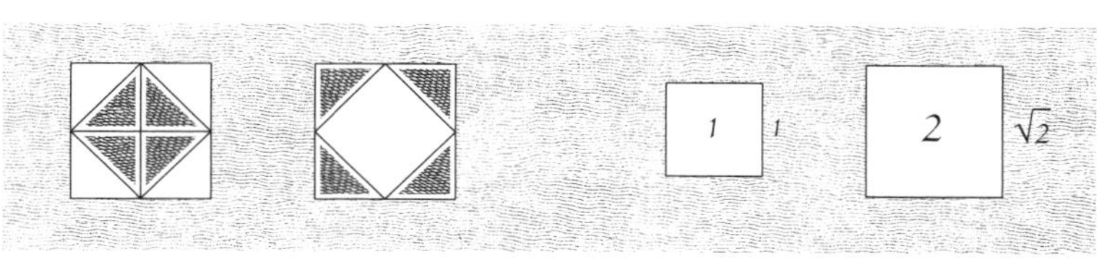

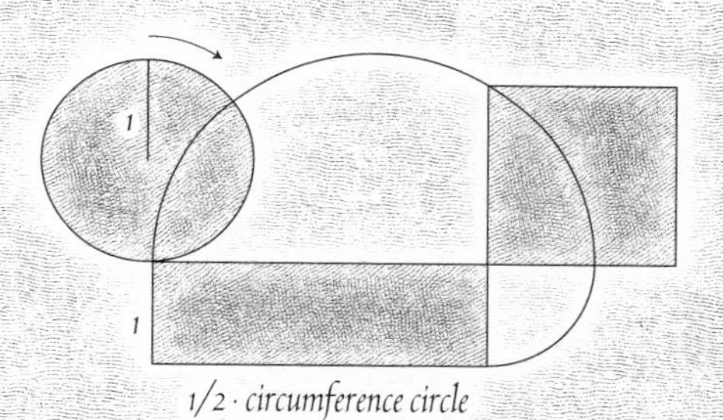

area circle = area rectangle = area square

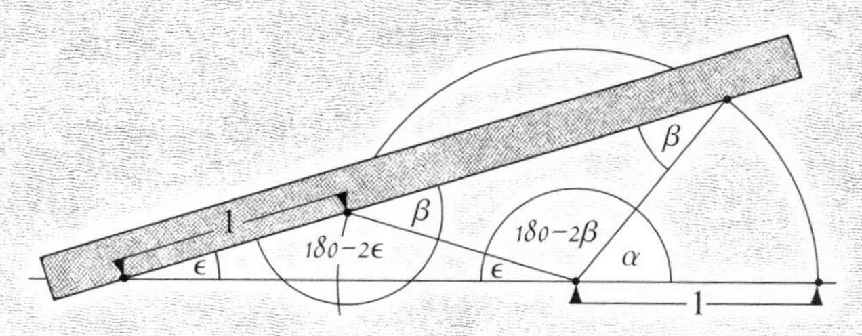

The first semicircle gives $\beta = 180 - (180 - 2\epsilon) = 2\epsilon$ *and the second* $\alpha = 180 - (\epsilon + (180 - 2\beta)) = 3\epsilon$.

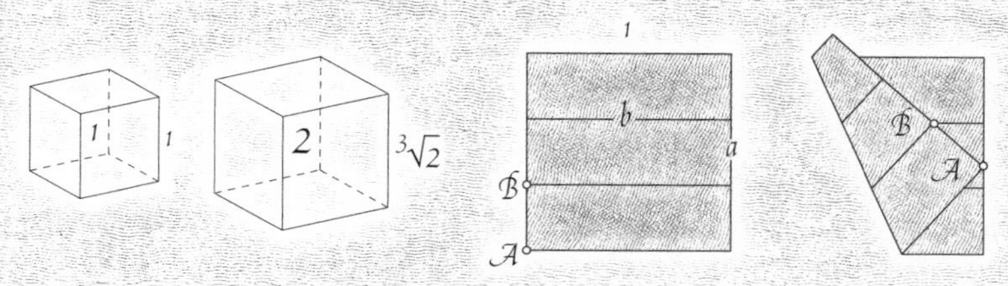

Folding $\mathcal{A}$ *and* $\mathcal{B}$ *onto the segments a and b results in* $\mathcal{A}$ *dividing the segment a in the ratio* $1:\sqrt[3]{2}$.

APPENDIX I

One Theorem, Many Proofs

Many hundred different proofs for Pythagoras's theorem (*see page 4*) have been discovered. Collected in this appendix are some of the most ingenious ones.

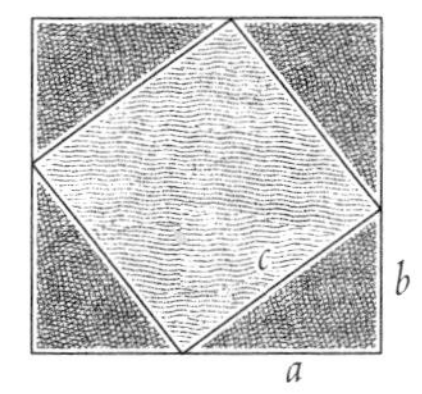

Our first proof (*left*) is based on the same diagram as the one on page 5:

$$\textit{big square} = \textit{small square} + 4\cdot\textit{triangle}$$
$$(a + b)^2 = c^2 + 4\cdot\tfrac{1}{2}ab$$
$$a^2 + b^2 + 2ab = c^2 + 2ab$$
$$a^2 + b^2 = c^2$$

A similar proof based on the diagram on the right:

$$\textit{small square} + 4\cdot\textit{triangle} = \textit{big square}$$
$$(b - a)^2 + 4\cdot\tfrac{1}{2}ab = c^2$$
$$b^2 - 2ab + a^2 + 2ab = c^2$$
$$a^2 + b^2 = c^2$$

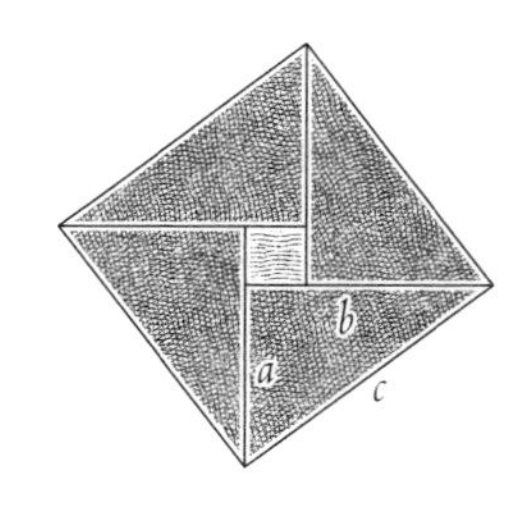

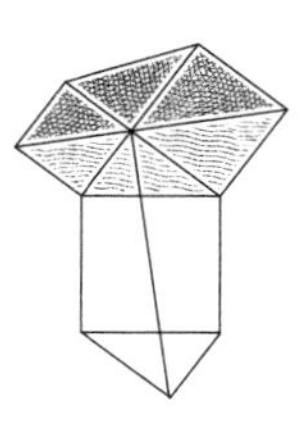

Leonardo da Vinci (1452–1519) noticed that the shaded areas in the two diagrams on the left have the same area and each contains two copies of the right-angled triangle. Discarding the triangles gives the theorem.

Triangles ABC, CBD, and ACD are similar. Thus, $DB/BC = BC/AB$ and $AD/AC = AC/AB$, or $BC^2 = AB\cdot DB$ and $AC^2 = AB\cdot AD$.
Sum up: $AC^2 + BC^2 = AB\cdot(DB + AD) = AB^2$

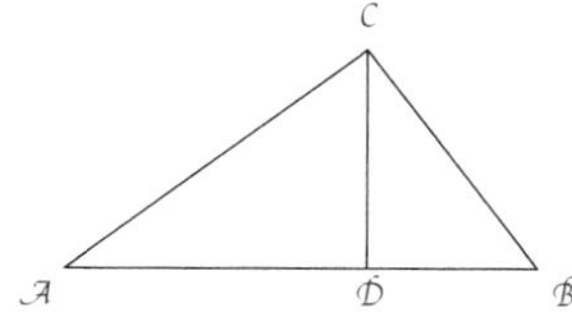

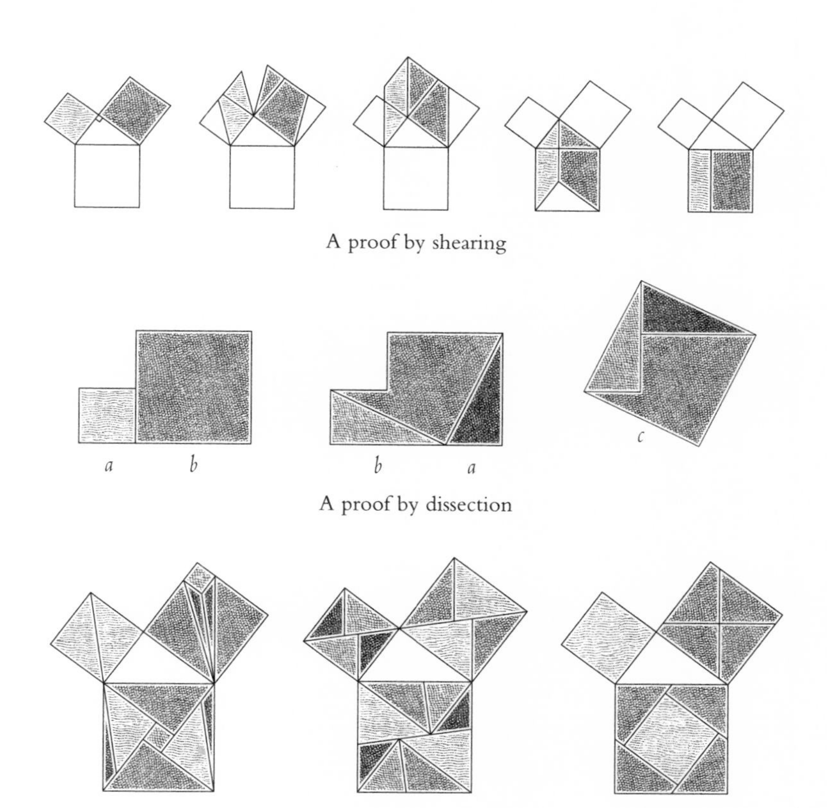

A proof by shearing

A proof by dissection

Three more proofs by dissection

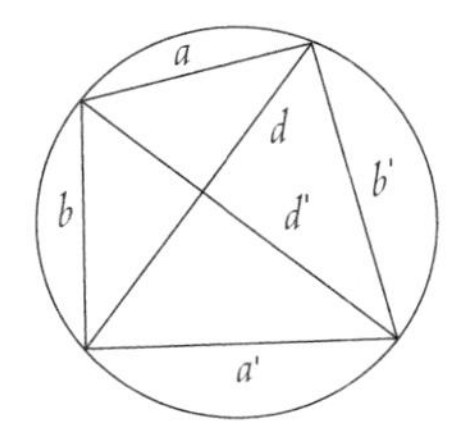

Ptolemy's theorem says that for a quadrilateral inscribed in a circle $a\,a' + b\,b' = d\,d'$. This reduces to Pythagoras's theorem if the quadrilateral is a rectangle.

APPENDIX II

ALL FOR ONE AND ONE FOR ALL

Delving deeply into the secrets of mathematics, one gets the feeling that everything is somehow connected to everything else by a network of beautiful relationships. For example, the Fibonacci numbers 1, 1, 2, 3, 5, . . . and the golden ratio $\phi = (\sqrt{5}+1)/2$ appear alongside many of the other topics covered in this little book. Connections with regular figures and Pascal's triangle have already been touched upon earlier. Here are some more in the theorem-proof style favored by mathematicians.

THEOREM 1:

$$\phi = 1+\cfrac{1}{1+\cfrac{1}{1+\cfrac{1}{1+\cfrac{1}{1+\cfrac{1}{\cdots}}}}}$$

PROOF: Call the infinite fraction x. Clearly, $x = 1 + 1/x$, or $x^2 - x - 1 = 0$. This equation has the solutions ϕ and $1-\phi$. Since $1-\phi$ is negative and both x and ϕ are not, x equals ϕ. Q.E.D.

Hence, just as 0.99... gets approximated by the numbers 0, 0.9, 0.99, . . . , ϕ gets approximated by the fractions

$$1\,, 1+\frac{1}{1}=\frac{2}{1}, 1+\frac{1}{1+\frac{1}{1}}=\frac{3}{2}, 1+\frac{1}{1+\frac{1}{1+\frac{1}{1}}}=\frac{5}{3}$$

THEOREM 2: *The nth fraction is* f_{n+1}/f_n *where* f_n *is the nth Fibonacci number.*

PROOF: The proof is by induction, as described in Mathematical Dominoes. Let's denote the nth fraction g_n. We first note that g_1, the first term of our sequence, is indeed $f_2/f_1 = 1/1 = 1$. Next, we assume that the nth term $g_n = f_{n+1}/f_n$. Then $g_{n+1} = 1 + 1/g_n = 1 + f_n/f_{n+1} = (f_{n+1} + f_n)/f_{n+1}$. With the defining equation of the Fibonacci numbers, $f_{n+1} + f_n = f_{n+2}$, this implies that $g_{n+1} = f_{n+2}/f_{n+1}$, as desired. Q.E.D.

THEOREM 3: *The Fibonacci numbers* f_n, $f_{n+1}, f_{n+2}, f_{n+3}$ *form the Pythagorean triple.*

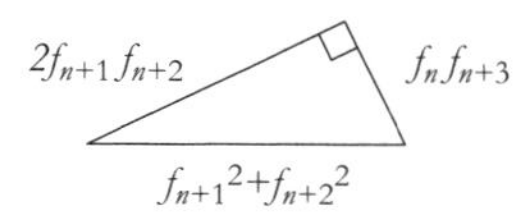

PROOF: If $a=f_n$, $b=f_{n+1}$, then $f_{n+2}=a+b$, $f_{n+3}=a+2b$ and $(2b(a+b))^2 + (a(a+2b))^2$ equals $(b^2+(a+b)^2)^2$. Q.E.D.

For example, choosing n = 1, 2, and 3 gives the triples 3 : 4 : 5, 5 : 12 : 13, and 16 : 30 : 34. From the proof it is clear that we can replace the Fibonacci numbers by any a, b, $a+b$, $a+2b$.

THEOREM 4: *ϕ is an irrational number.*

PROOF: As in The Nature of Numbers, we assume that ϕ is a fraction a/b of two natural numbers.

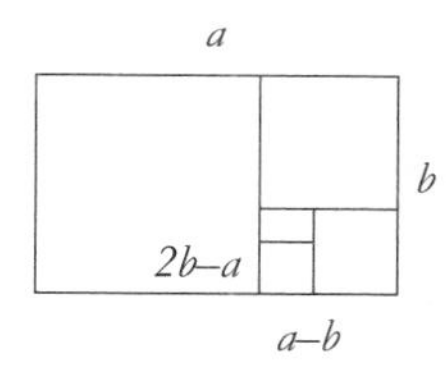

Thus we have a golden rectangle with side lengths a and b. Using the self-similarity property of golden rectangles (*see page 40*), we cut off a square to produce another golden rectangle. Repeating this for the smaller golden rectangle produces a third, and so on. We notice that the side lengths of these golden rectangles form the infinitely decreasing sequence of natural numbers $a > b > a-b > 2b-a > \ldots$. Since any decreasing sequence of natural numbers must end (1 being the lower limit), this is a contradiction to our assumption. Hence ϕ is irrational. Q.E.D.

The infinite expression

$$\sqrt{2+\sqrt{2+\sqrt{2+\sqrt{\ldots}}}}$$

is the main part of our formula for the value of π (*see page 9, bottom*).

THEOREM 5:

$$\phi = \sqrt{1+\sqrt{1+\sqrt{1+\sqrt{\ldots}}}}$$

PROOF: Call the infinite expression y. Then, because of its self-similarity $y = \sqrt{1+y}$, or $y^2 - y - 1 = 0$. Thus $y = \phi$. Q.E.D.

We finish with another beautiful connection between the Fibonacci numbers and ϕ.

THEOREM 6: *The nth Fibonacci number is*

$$f_n = (\phi^n - (1-\phi)^n)/(2\phi - 1).$$

PROOF: In the proof of Theorem 2, we saw that both ϕ and $\tau = 1-\phi$ satisfy the equation $x^2 = x+1$. We conclude that $\phi^2 = 1\phi+1$, $\phi^3 = \phi^2+\phi = 2\phi+1$, $\phi^4 = \phi^3+\phi^2 = 3\phi+2$. It follows by induction that $\phi^n = f_n\phi + f_{n-1}$ and, similarly, $\tau^n = f_n\tau + f_{n-1}$. Subtracting the second equation from the first gives $\phi^n - \tau^n = f_n(\phi - \tau) = f_n(2\phi - 1)$. Finally, dividing both sides of this equation by $2\phi - 1$ gives the desired formula. Q.E.D.

Since $(1-\phi)^n = (-0.6180\ldots)^n$ is tiny, the nth Fibonacci number is the closest natural number to $\phi^n/(2\phi-1)$. For example, $\phi^{10}/(2\phi-1) = 55.0036\ldots$ and the 10th Fibonacci number is 55.

APPENDIX III

LOOKS CAN BE DECEIVING

Some of the proofs in this collection, especially the dissection ones and those involving infinity, omit a lot of detail and really only try to capture the essence of why something is true. Many of them would only count as full proofs in the eyes of a mathematician if more details were added. Here are a number of infamous fallacies that use arguments that are very similar to some used in this book.

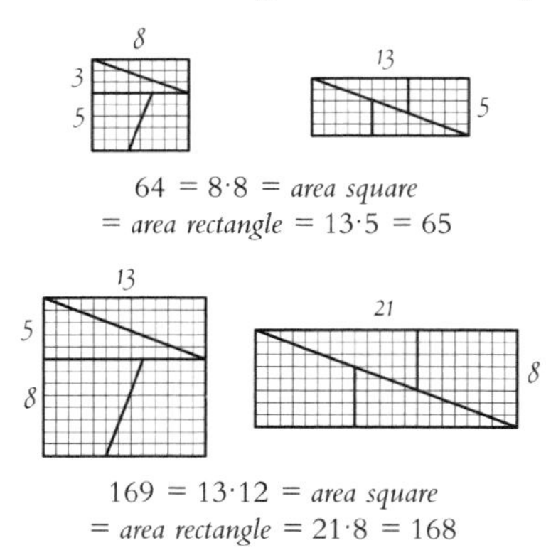

$64 = 8 \cdot 8 =$ *area square*
$=$ *area rectangle* $= 13 \cdot 5 = 65$

$169 = 13 \cdot 12 =$ *area square*
$=$ *area rectangle* $= 21 \cdot 8 = 168$

DISSECTION DISASTER: The first "proof" by dissection shows that $64 = 65$. What goes badly wrong here is that the diagonal cut in the rectangle on the right is not really a line as suggested by the drawing, but a very thin quadrilateral slit of area one. This "proof" as well as the next is based on the fact that the square on any Fibonacci number differs by one from the product of its two neighbors. In the first "proof" the square has a smaller area than the oblong rectangle, in the second it is the other way around.

INFINITE INSANITY: In From Pie to Pi we argued that the regular n-gons inscribed in a circle approximate it in shape, and that hence their circumferences approach that of the circle. This is true but requires a proof, as the following fallacy shows.

Starting with the large semicircle on the right, its diameter is approximated by strings of ever smaller semicircles. Hence the lengths of the strings approximate that of the diameter. However, every string is clearly exactly as long as the large semicircle. Hence, a semicircle is just as long as its diameter, or $\pi = 1$.

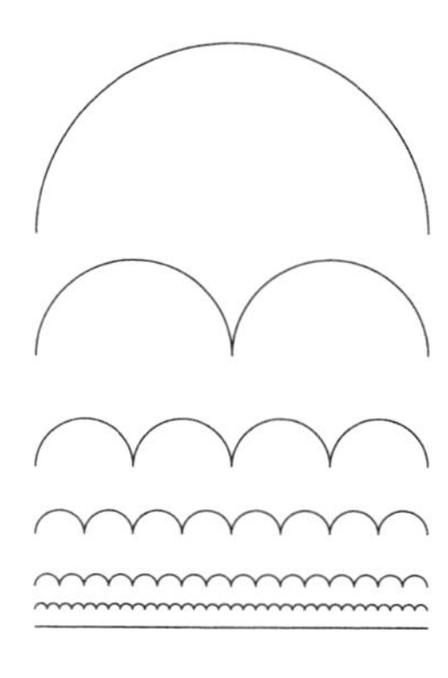

ROUTINE RISK: In The Infinite Staircase we manipulated the following infinite sum as we would manipulate a finite one to prove that it adds up to infinity.

$$1 + \frac{1}{2} + \frac{1}{3} + \frac{1}{4} + \frac{1}{5} + \frac{1}{6} + \frac{1}{7} + \frac{1}{8} + \frac{1}{9} + \frac{1}{10} + \frac{1}{11} + \frac{1}{12} + \cdots$$

We should be careful when doing this. For example, the following similar sum adds up, in a precise sense, to 0.6931... = ln 2, the natural logarithm of 2.

$$1 - \frac{1}{2} + \frac{1}{3} - \frac{1}{4} + \frac{1}{5} - \frac{1}{6} + \frac{1}{7} - \frac{1}{8} + \frac{1}{9} - \frac{1}{10} + \frac{1}{11} - \frac{1}{12} + \cdots$$

By just rearranging this sum, we can "prove" that ln 2 = ½ (ln 2), or 2 = 1.

$$\begin{aligned}
&= (1 - \tfrac{1}{2}) - \tfrac{1}{4} + (\tfrac{1}{3} - \tfrac{1}{6}) - \tfrac{1}{8} + (\tfrac{1}{5} - \tfrac{1}{10}) - \tfrac{1}{12} + (\tfrac{1}{7} - \tfrac{1}{14}) - \cdots \\
&= \tfrac{1}{2} - \tfrac{1}{4} + \tfrac{1}{6} - \tfrac{1}{8} + \tfrac{1}{10} - \tfrac{1}{12} + \tfrac{1}{14} - \cdots \\
&= \tfrac{1}{2}(1 - \tfrac{1}{2} + \tfrac{1}{3} - \tfrac{1}{4} + \tfrac{1}{5} - \tfrac{1}{6} + \tfrac{1}{7} - \cdots
\end{aligned}$$

Even worse, it can be proved that for any real number, there is a rearrangement of this sum into a sum that adds up to this number. Of course, all this does not mean that you cannot do anything meaningful with infinite expressions, only that you have to follow certain rules.

UNIVERSAL UNIQUENESS: Many proofs that there are exactly five regular solids start, as we did in our introduction, by showing that there are essentially five possible corners, and end by constructing five solids with these kinds of corners. These proofs are incomplete because they don't show the uniqueness of these solids. Just imagine building a hollow icosahedron such that adjacent rigid faces meet at their edges in hinges. How can you be sure that what you end up with is rigid? After all, in isolation every one of the corners can be flexed, so why not the whole shape?

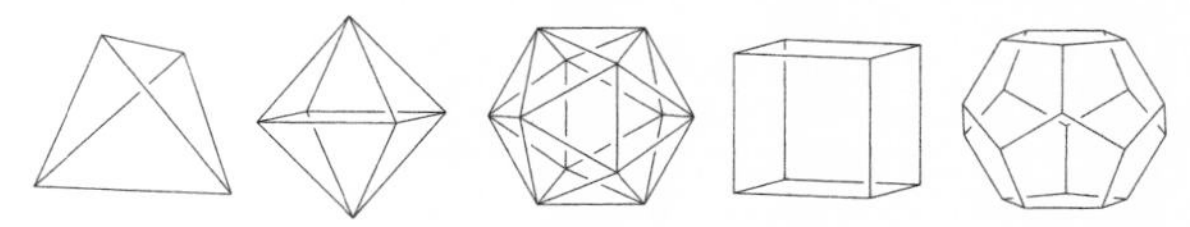

Here is a related question: If you build the skeletons of the regular solids using rigid edges only, such that adjacent edges can be flexed about a common vertex, then which of the skeletons are rigid and which are not?

APPENDIX IV

Triangles of Generality

Deep results are not usually discovered at-a-glance but rather as a result of a process of ever-increasing generalization. In this appendix, we sketch part of such a process involving Pascal's triangle and its relative, the Power Triangle.

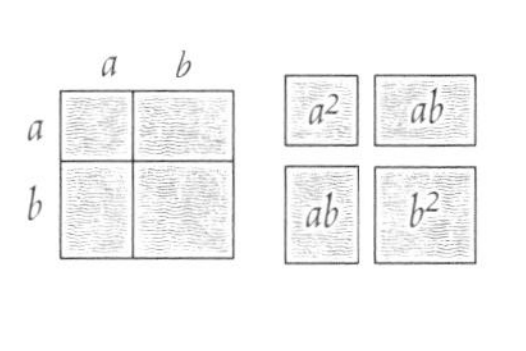

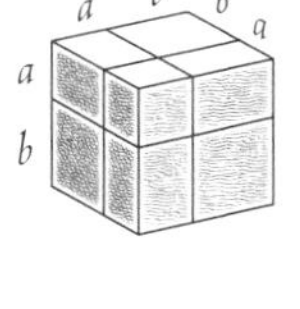

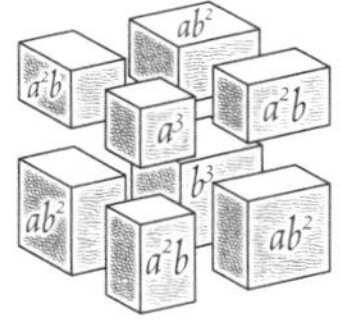

$$(a+b)^2 = 1\,a^2 + 2\,a\,b + 1\,b^2 \qquad (a+b)^3 = 1\,a^3 + 3\,a^2\,b + 3\,a\,b^2 + 1\,b^3$$

These general formulas work for any choice of *a* and *b*. Here is how you derive a general formula for $(a+b)^n$ that works for any *a* and *b* and natural number *n*.

$$\begin{array}{ccccccccc} &&&&1&&&& \\ &&&1&&1&&& \\ &&1&&2&&1&& \\ &1&&3&&3&&1& \\ 1&&4&&6&&4&&1 \end{array}$$

$$(a+b)^0 = 1$$
$$(a+b)^1 = 1\,a + 1\,b$$
$$(a+b)^2 = 1\,a^2 + 2\,a\,b + 1\,b^2$$
$$(a+b)^3 = 1\,a^3 + 3\,a^2\,b + 3\,a\,b^2 + 1\,b^3$$
$$(a+b)^4 = 1\,a^4 + 4\,a^3\,b + 6\,a^2\,b^2 + 4\,a\,b^3 + 1\,b^4$$

Pascal's triangle on the left summarizes the formulas on the right. Can you prove that each of its entries (except the tip) is the sum of the numbers right above it?

$$\begin{array}{ccccccccc} &&&&\binom{0}{0}&&&& \\ &&&\binom{1}{0}&&\binom{1}{1}&&& \\ &&\binom{2}{0}&&\binom{2}{1}&&\binom{2}{2}&& \\ &\binom{3}{0}&&\binom{3}{1}&&\binom{3}{2}&&\binom{3}{3}& \\ \binom{4}{0}&&\binom{4}{1}&&\binom{4}{2}&&\binom{4}{3}&&\binom{4}{4} \end{array}$$

On page 21, we saw that Pascal's triangle coincides with the triangle on the left. Here $\binom{n}{k}$ is the number of different ways to choose *k* objects among *n* objects, which is 1 if *k* is zero and otherwise is $n \cdot (n-1) \cdot (n-2) \ldots (n-k+1) / (1 \cdot 2 \cdot 3 \ldots k)$. This gives the famous Binomial Theorem:

$$(a+b)^n = \binom{n}{0}a^n + \binom{n}{1}a^{n-1}b + \binom{n}{2}a^{n-2}b^2 + \ldots + \binom{n}{n}b^n$$

Many generalities hide in Pascal's triangle. For example, on page 43 we saw that its nth diagonal adds up to the nth Fibonacci number f_n. We express this as follows:

$$f_n = \binom{n-1}{0} + \binom{n-2}{1} + \binom{n-3}{2} + \ldots$$

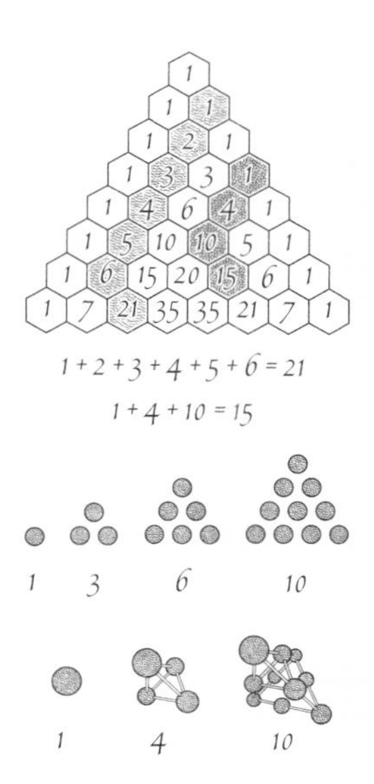

The structure of Pascal's triangle quickly suggests the Golf Club Theorem: Highlight a golf-club-shaped pattern of numbers in the triangle as in the examples on the left. The numbers in the handle of the club then add up to the one in its tip! Golf clubs with handles along the outside column on the left (first, white) translate into the formula:

$$1 + 1 + 1 + \ldots + 1 \; (n \text{ times}) = \binom{n}{1} = n$$

Golf clubs with handles in the second column translate into the formula for the sum of the first n natural numbers (*see also pages 32 and 34*):

$$1 + 2 + 3 + \ldots + n = \binom{n+1}{2} = n(n+1)/2$$

These sums are called *triangular numbers* because they count the numbers of circles in the triangular patterns on the left. Similarly, it follows that the numbers in the next column are *pyramidal numbers*. In general, the numbers in column n of Pascal's triangle are the (n–1)-dimensional pyramidal numbers.

The Power Triangle below was only recently discovered. It summarizes the general formulas on the right below, some of which we proved on page 34. This triangle grows just like Pascal's except that you have to multiply a number by its suffix before adding it. For example, in the fourth row: 7·2 + 12·3 = 50 and 12·3 + 6·4 = 60.

1_1	$1^0 + 2^0 + 3^0 + \ldots + n^0 = 1\binom{n}{1}$
$1_1 \;\; 1_2$	$1^1 + 2^1 + 3^1 + \ldots + n^1 = 1\binom{n}{1} + 1\binom{n}{2}$
$1_1 \;\; 3_2 \;\; 2_3$	$1^2 + 2^2 + 3^2 + \ldots + n^2 = 1\binom{n}{1} + 3\binom{n}{2} + 2\binom{n}{3}$
$1_1 \;\; 7_2 \;\; 12_3 \;\; 6_4$	$1^3 + 2^3 + 3^3 + \ldots + n^3 = 1\binom{n}{1} + 7\binom{n}{2} + 12\binom{n}{3} + 6\binom{n}{4}$
$1_1 \;\; 15_2 \;\; 50_3 \;\; 60_4 \;\; 24_5$	$1^4 + 2^4 + 3^4 + \ldots + n^4 = 1\binom{n}{1} + 15\binom{n}{2} + 50\binom{n}{3} + 60\binom{n}{4} \ldots$

APPENDIX V
POLYTOPES OF ANALOGY

This book started with the regular two- and three-dimensional polytopes, the regular polygons, and Platonic solids. We end it by showing how analogy can be used to guess and prove the properties of higher-dimensional regular polytopes.

BASIC POLYTOPES: Using coordinates, it is easy to see that the tetrahedron, cube, and octahedron have the following relatives in n-dimensional (or n-d) space.

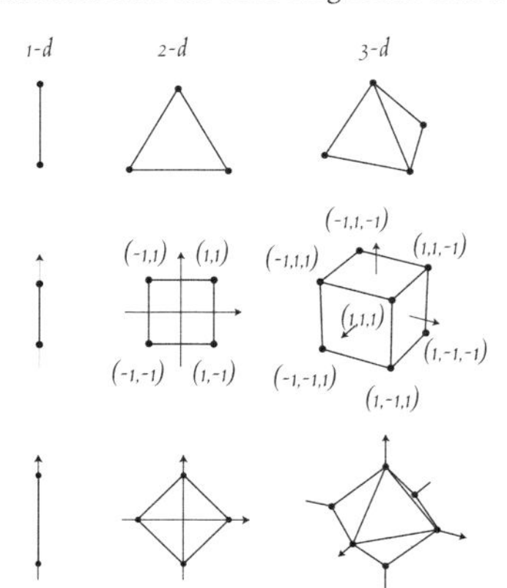

The n-d *simplex* has n+1 vertices, two of which are the same distance apart. A 1-, 2-, 3-, . . . , n-d simplex is bordered by 2, 3, 4, . . . , n+1 simplices of one less dimension.

The vertices of an n-d *cube* are the points with all coordinates 1 or -1. A 1-, 2-, 3-, . . . , n-d cube has 2, 4, 8, . . . , 2^n vertices and is bordered by $2n$ cubes of one less dimension. The *tesseract* is the 4-d cube.

The vertices of the n-d *orthoplex* are the $2n$ end points of an n-d unit cross. A 1-, 2-, 3-, . . . , n-d orthoplex has 2, 4, 6, . . . , $2n$ vertices and is bordered by 2, 4, 8, . . . , 2^n simplices of one less dimension.

CLASSIFICATION: Ludwig Schlafli (1814–1895) proved that apart from the simplex, cube, and orthoplex, there are three more regular polytopes in four dimensions and no more in any higher dimension.

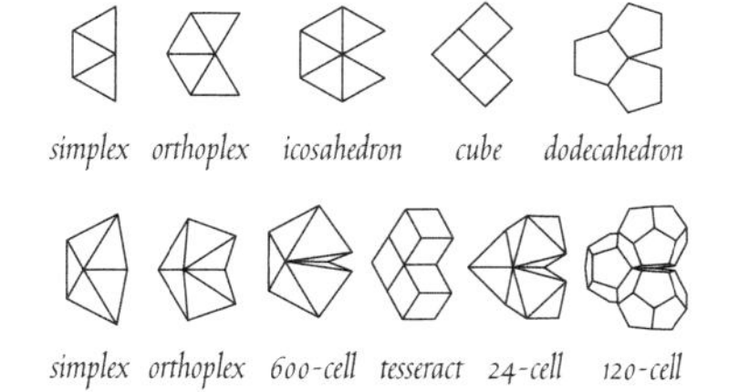

The regular 3-d polytopes correspond to the five ways of fitting at least three identical regular 2-d polytopes around a vertex with space left to fold up into the third dimension (*see page vi*). The regular 4-d polytopes correspond to the six ways of fitting at least three identical regular 3-d polytopes around an edge with space left to fold up into 4-d space.

VITAL STATISTICS: The regular 3-d polytopes have identical *vertices*, 1-d *edges*, and 2-d *faces*. The regular 4-d polytopes also have 3-d *cells*. The numbers V (vertices), E (edges), F (faces), C (cells), f/v (faces per vertex), c/e (cells per edge), and c/v (cells per vertex) are:

3-d polytope	faces	F	E	V	f/v
tetrahedron	**triangle**	**4**	**6**	**4**	**3**
cube	**square**	**6**	**12**	**8**	**3**
octahedron	**triangle**	**8**	**12**	**6**	**4**
icosahedron	**triangle**	**20**	**30**	**12**	**5**
dodecahedron	**pentagon**	**12**	**30**	**20**	**3**

4-d polytope	cells	C	F	E	V	c/e	c/v
simplex	**tetrahedron**	**5**	10	10	**5**	**3**	4
tesseract	**cube**	**8**	24	32	**16**	**3**	4
orthoplex	**tetrahedron**	**16**	32	24	**8**	**4**	8
24-cell	**octahedron**	24	96	96	24	**3**	6
120-Cell	**dodecahedron**	120	720	1200	600	**3**	4
600-Cell	**tetrahedron**	600	1200	720	120	**5**	20

The bold entries are easily extracted from the information on the opposite page and, in the case of the 3-d polytopes, from some life models. Here is how you extract the numbers c/v (cells per vertex) for the 4-d polytopes by analogy.

Cutting a regular 3-d polytope close to a vertex gives a regular polygon whose sides are the cuts of the faces around the vertex. Thus, *no. sides of polygon* = f/v *of 3-d polytope*

Cutting a regular 4-d polytope close to a vertex gives a regular 3-d polytope whose faces are the cuts of the cells around the vertex. Hence,

no. faces of 3-d polytope = c/v *of 4-d polytope*

f/v *of 3-d polytope* = c/e *of 4-d polytope*

These relationships allow one to deduce the cuts of the different 4-d polytopes on the right and thereby also their c/v. So, for example, since the 4-d cube has a tetrahedral cut, it has four cells per vertex.

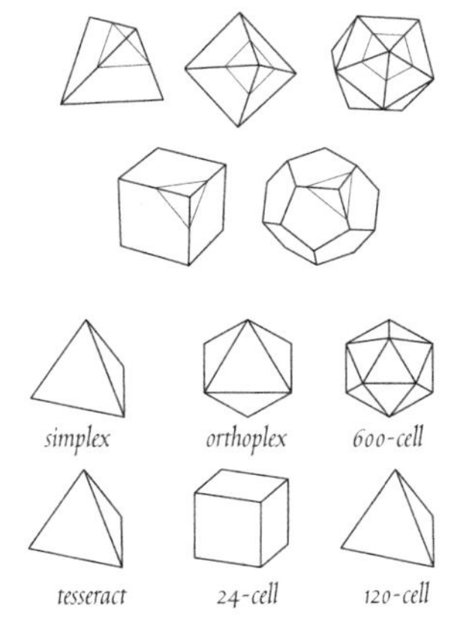

The numbers of cells of the three nonstandard regular 4-d polytopes are built into their names, and no one-glance way to deduce these numbers has been discovered. However, once we know these numbers, the part of the 4-d table that we have not yet touched can be easily filled in using some simple relationships such as: (1) the 4-d analogue of the Euler formula $V+F-E-C=0$ (*see page 44*), (2) $F = C \cdot$ faces per cell/2, and last but not least, (3) $E = C \cdot$ edges per cell/cells per edge.

VISUALIZATION: Take one of the Platonic solids, choose a point outside but very close to the center of one of the faces, and from this point project the skeleton of the solid onto the face. The resulting 2-d image contains most of the information about the solid.

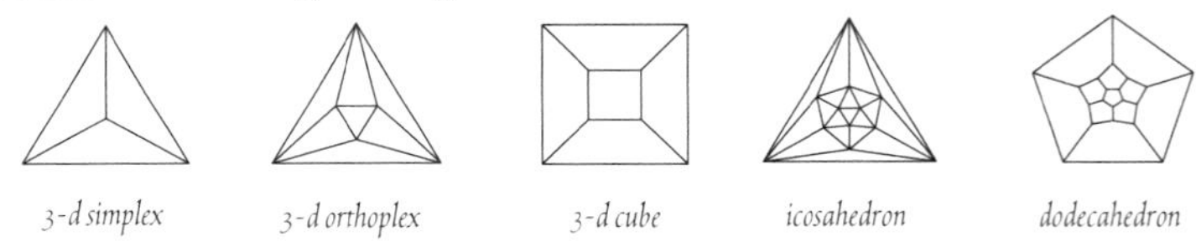

3-d simplex *3-d orthoplex* *3-d cube* *icosahedron* *dodecahedron*

By performing the analogous operation for the regular 4-d polytopes, we arrive at the following 3-d images. Note that the first three images above generalize to those below. The projections of the 120- and 600-cell are too complex to be reproduced here.

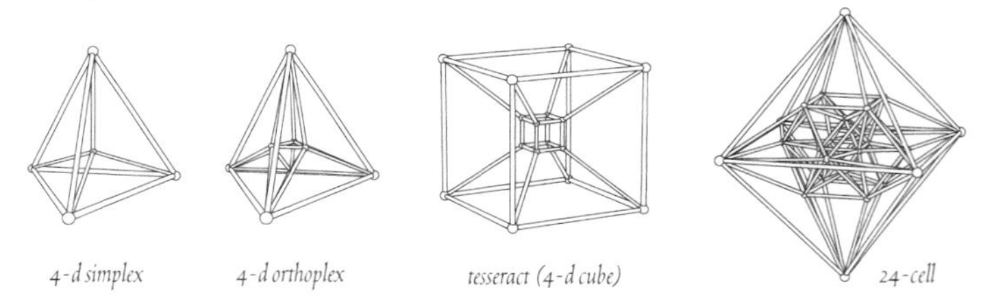

4-d simplex *4-d orthoplex* *tesseract (4-d cube)* *24-cell*

CONSTRUCTION: Constructing the *n*-d simplex, orthoplex, and cube is easy, and we can make the other regular polytopes from these standard ones. On page 41 we constructed the icosahedron from the octahedron and a dodecahedron can be inscribed as a *dual* in the icosahedron (*below, right*) such that the midpoints of the faces of the second are the vertices of the first. This gives all the Platonic solids.

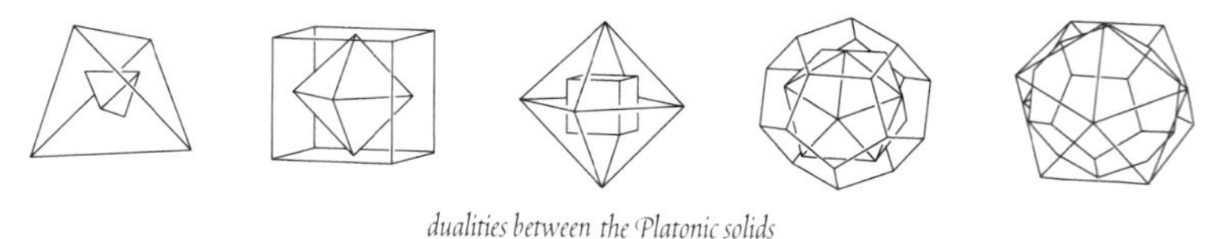

dualities between the Platonic solids

The midpoints of the faces of a 4-d cube are the vertices of a 24-cell. By dissecting its octahedral cells as described on page 41, we get 96 points and 24 icosahedra. For each such icosahedron there is a point in 4-d at the same distance from all its vertices as the icosahedral edge length. Then the 24 points corresponding to the icosahedra plus the 96 points are the vertices of a 600-cell. Finally, a 120-cell can be inscribed as a dual into the 600-cell (the midpoints of the cells of the second are the vertices of the first).